i

CONTENTS

CHAPTER 3

McCook Field, March 1919 - October 1927

PREFACE

This book was written in support of UES, Inc. contract #F33615-90C-2086, Task 21, in accordance with WL/POTF requirements. Mr. Marv Stibich of Wright Laboratory supervised the project. Additional volumes of Aero Propulsion and Power Directorate documentation are expected.

The *McCook Field Years* details the early history of flight, from the invention of ballooning to the Wright B Flyer and beyond. It chronicles the progression of the United States Air Corps and its significance before, during, and after World War I. Most importantly, the book focuses on the formation of McCook Field in Dayton, Ohio and the critical role it played in the advancement of aeronautics in the United States.

The book was written by Mr. Albert Misenko, assisted by Mr. Robert Twist, and edited by Ms. Rose Strick under supervision of Project Manager Mr. James St. Peter, UES Historian. Ms. Strick also prepared the document for publication, including the collection of McCook Field photographs. The United States Air Force Museum Research Center and the UES History Office in Bldg. 435, Area B, Wright-Patterson Air Force Base, supplied all photographs featured.

PROLOGUE

MASTERING THE AIR, 1783 - 1909

Balloons and Dirigibles

Human flight began with a gas bag.[1] In 1783, brothers Jacques and Joseph Montgolfier of France launched the earliest balloons. Interestingly enough, the first passengers were not humans, but rather a sheep, a rooster, and a duck. They landed safely. In October, the first person, Jean Francois Pilatre de Rozier, ventured aloft in a basket suspended below a gas bag tethered to the ground. He found the higher atmosphere was not exceptionally hostile so he and the Marquis d'Arlandes launched themselves into the sky above Paris. Balloonists, propelled by nature's winds, had an inexpensive source of energy but one they could not readily harness.

In December 1785, Jean-Pierre Blanchard and Dr. John Jeffries made the first balloon crossing of the Channel, flying from England to France. Blanchard's flight underlined the military potential of balloons and in 1794 the French organized a balloon corps. Napoleon dismissed the corps in 1798 despite the successful use of balloonists for observation against the Austrians on the battlefield.

Meanwhile, Blanchard later demonstrated his balloon in the United States, flying from Philadelphia into New Jersey. In the crowd that witnessed his ascent was the new president of the United States, George Washington, and several of his successors such as John Adams, Thomas Jefferson, James Madison, and James Monroe. Ballooning quickly became a spectator sport in the U.S. during the first half of the nineteenth century.

Ballooning in the U.S. for military purposes, however, did not occur until the Civil War.[2] Urged on by President Abraham Lincoln, the Union Army established a balloon corps in 1861 as the Army's first air arm. Using mobile hydrogen generators, the corps operated several balloons either tethered to the ground or to a boat. Observers perched in baskets suspended below the balloons telegraphed or dropped messages to the ground.

Success in directing artillery fire and reporting troop movements could not overcome the skepticism of the North's officers, and the balloon corps disbanded in June 1863. The Confederates, though appreciative of the Union's observation advantage against them, lacked funds for balloons and therefore used them even more briefly. Despite formation of balloon corps in several European armies, the U.S. Army did not use them again until the Spanish-American War. Thus, between 1863 and 1890, military balloon operations in the U.S. did not exist.

In October 1890, Congress assigned the Signal Corps to collect and transmit information for the Army. The Chief Signal Officer regarded it as including aerial operations, and in 1891 he requested appropriations to develop a balloon corps. In 1892, the Chief Signal Officer established a balloon section.

When the war with Spain began in 1898, the Signal Corps shipped its one balloon to Cuba for observation and fire-control operations. Unfortunately, the mission ended when bullets punctured the balloon shortly after its arrival. Balloons were not used again until 1902 when a balloon detachment was organized at Fort Myer. Still, the Army participated little in ballooning during the next few years due to a lack of balloonists and other resources.

Meanwhile in France, Henri Giffard in 1852 used a lightweight steam engine to fly the first dirigible, a steerable airship powered by an engine. The motor was a 100-pound, 1-cylinder steam engine driving a 3-bladed propeller. Steam engines proved too heavy to power balloons efficiently, delivering only 1 horsepower for 112 pounds of weight. In 1885, Karl Benz, a German engineer, demonstrated the first practicable automobile driven by a gasoline engine. Development of gasoline engines for automobiles preceeded their application to dirigibles by only a few years. The first to do so was a wealthy Brazilian living in Paris, Alberto Santos-Dumont, in 1898. He adapted an automobile engine to his airship by attaching a propeller to the drive shaft. The pilots' seat and other necessary equipment stood mounted below the propeller. He flew a number of airships before turning to airplanes.

The advent of sport ballooning in Europe inspired the formation of the Aero Club of America in 1905. As a member of the Federation Aeronautique Internationale, an organization that set international standards for balloon, dirigible, and airplane pilots, it played a major role in advancing aeronautics in America.

The U.S. Army, however, failed to keep up with the accelerated aeronautical activity in Europe after 1900. The balloon detachment of the Signal Corps was virtually non-existent after 1898. The Signal Corps had acquired only ten balloons from the end of the Civil War through 1907. On August 1, 1907, the Signal Corps established an Aeronautical Division with Captain Charles deForest Chandler in charge. The new division, forerunner of the Air Force, had control of all aeronautical matters and related subjects.

Before and after the Spanish-American War, the Signal Corps called for purchasing a dirigible. Dirigibles came in two classes: rigid ones maintained the envelope with a rigid structure; nonrigid ones kept the envelope using pressure from the gas inside. Military services abroad were using dirigibles for reconnaissance and transporting officers and couriers. Some thought that dirigibles could serve as a venue for bombing also.

Exhibitions showcased dirigibles in the United States in the early 1900s, earning prize money for altitude, endurance, distance, and time trials. Thomas S. Baldwin was one of the private persons in the U.S. constructing and flying small airships at fairs. He initially adapted one of his motorcycle engines for use in a dirigible, but eventually sought help from Glenn H. Curtiss who designed and produced 2-cylinder engines for his use.

The Signal Corps did not receive money to purchase a dirigible until 1908. At that time, Baldwin contracted with the Chief Signal Officer to build an airship for the Army. He delivered the dirigible to Fort Myer and equipped it with a new lightweight Curtiss engine specially designed for it. The 4-cylinder, water-cooled engine, arranged in a vertical line, generated between 20 and 30 horsepower. The engine, mounted one-third of the way from the car's front, drove a tubular steel shaft with a single wooden propeller designed by Lt. Thomas E. Selfridge.

The airship was speed tested on August 14, 1908. It averaged less than the 20 miles per hour required by the specifications. Thus, the original price of $6,750 dropped 15 percent. The Signal Corps formally accepted the airship on August 28, 1908 as Signal Corps Dirigible Number 1. The dirigible was used for instruction and exhibition. Baldwin trained Lieutenants Frank P. Lahm, Benjamin D. Foulois, and Selfridge to pilot the craft. On May 26, 1909, Lahm and Foulois made their first ascent, and became the Army's first pilots.

Gliders and Airplanes

Gliders made thousands of short flights beginning early in the nineteenth century. Sir George Cayley, regarded as the inventor of the airplane and founder of the science of aerodynamics, was one of Europe's leading scientists active in aeronautical research. Before his death in 1857, Cayley designed and flew many gliders, including a full-size one capable of carrying a person.

During the middle and late nineteenth century, other experimenters constructed flying craft based on Cayley's work. These designers also tried to find an efficient, lightweight power plant. Like balloonists, they found steam engines too heavy and cumbersome for flying. Otto Lilienthal, a German aeronautical engineer in the 1890s, marked another milestone in flight when he became the first person to launch himself into the air and fly consistently in heavier-than-air craft. He made thousands of successful flights in his hang gliders, including the use of biplane wings, before he died following a crash in 1896. His flights, publicized around the world in story and photographs, contributed vast knowledge about materials, design, construction, and control of gliders.

In the United States, the French-born engineer Octave Chanute acted as both researcher and publicist. He published Progress in Flying Machines in 1894, a history and reference work. Chanute's biplane glider of 1896 offered an improved method of rigging the wings. Contacted by the Wrights in 1900, he thereafter encouraged their work and publicized their successes until his death in 1910.

As collections of aeronautical data accumulated, many other aviation pioneers used the information to advance flying. Samuel Pierpont Langley, astronomer and head of the Smithsonian Institution in

Washington, D.C., attempted to build a powered airplane piloted by a human. Langley tested his design for the <u>Aerodrome</u> in a series of gliders. He also flew models powered by small steam engines and by carbonic-acid gas and air in the 1890s. He even built and flew a quarter-scale version of his full-size aircraft. Encouraged by President William McKinley, the assistant secretary of the Navy, Theodore Roosevelt, helped Langley secure a $50,000 grant from the War Department in 1898.

Finding steam engines too heavy for practicable use, Langley equipped his full-size airplane with a gasoline engine. He placed two pusher counter-rotating propellers located at either side of the fuselage behind the front wings. The blades, consisting of a triangular framework of tubes covered with fabric, were 2 feet wide at their outer ends. At the rear of the nacelle, which occupied the space between the lower engine section and the forward quadrilateral frame, was the large carburetor for the engine.

Charles M. Manly perfected the gasoline engine used by Langley based on Stephen M. Balzar's earlier work on an air-cooled rotary.[3] Manly's 5-cylinder, fixed-radial, water-cooled engine weighed only 125 pounds (207.5 pounds with 20 pounds of cooling water, batteries, and accessories) and produced 52.4 horsepower. It was one of the most efficient gasoline engines produced up to World War I. Unfortunately, Manly crashed twice trying to fly the Langley airplane in 1903, without ever having made so much as a single glider flight.. The problem was not in Manly's engine but in Langley's catapult system, the airplane's structure, and the control system. Subsequent criticism of the Army's Board of Ordnance and Fortification made it leery of funding flying machines.

Meanwhile, the Wright brothers in Dayton, Ohio, had independently developed a successful aircraft.[4] Although they lacked in professional training, their intelligence and mechanical aptitude transformed them into outstanding engineers and inventors. Their interest in flying arose after witnessing Lilienthal's glider flights. In May 1899, they wrote to the Smithsonian for literature on aviation, and they subsequently benefited from relations with Chanute who was in contact with aeronautical research in Europe and the United States. They built a biplane kite in August 1899. Thereafter, the Wrights compiled new data on aerodynamics, designing their own wind tunnel in 1901, and built their own gliders.

At Kitty Hawk in North Carolina, the brothers gained experience in controlled flying with their gliders beginning in 1900. By 1902, the brothers had solved the problem of control and had become skillful glider pilots. Unable to acquire a suitable lightweight motor, the Wrights built their own engine.[5] Machinist Charles E. Taylor assisted in the project.. Their efforts produced a 4-cylinder, water-cooled engine with a 4-inch bore and stroke that developed 12 horsepower. The engine weighed 161 pounds dry, but a little over 200 pounds with magneto, radiator, tank, tubing, accessories, water, and fuel. They also had to perfect an aeronautical propeller because the published material on the subject was useless to them. They mounted two propellers, rotating in opposite directions, behind the wings of their aircraft. On December 17, 1903 at Kill Devil Hill, the brothers made the first controlled flights in a powered airplane.

Early in 1904, the Wrights constructed a second airplane and engine, a much more efficient one of 15-16 horsepower. They also relocated their flying experiments to Huffman Prairie, about eight miles from Dayton. During these experiments, the brothers continued to learn about flying. They constructed a derrick catapult for launching the aircraft and made a number of changes in the machine itself. By year's end, they had made two five-minute flights. In May 1905, they returned to the prairie with a new machine, the first practicable aircraft. They used the 1904 engine but developed new propellers. The aircraft made a flight in October 1905 of more than 24 miles. A need for fuel ended the 38 minute flight. The brothers discontinued their flying activities, fearing that their machine's design was becoming public knowledge. In 1906, the government granted them a patent.

Both the public and the war department ignored the Wrights despite their success in 1903 and their progress toward perfecting the airplane at Huffman Prairie in 1904 and 1905. The brothers negotiated with Great Britain, France, and Germany before securing a contract with the War Department in February 1908. Theodore Roosevelt, now president, directed the War Department to investigate the Wright brothers' claims.

In preparation for their demonstration flights, the brothers returned to Kitty Hawk in May 1908 after having not flown for 30 months. They modified the 1905 Flyer with seats for two persons and a new engine equipped with 4 vertically-placed cylinders, more than twice as powerful as the 1903 engine. Wilbur's flights in Europe and Orville's tests at Fort Myer, Virginia, in 1908 and 1909, brought them world recognition. Orville's flight with Lt. Thomas E. Selfridge in September 1908 ended in tragedy when a crack in the right propeller caused it to loosen and foul a rudder guy wire. Both broke and the aircraft crashed. Selfridge, fatally injured, died a few hours later. Orville, seriously injured, remained hospitalized for several weeks.

In June 1909, the Wrights returned to Fort Myer to resume their aircraft trials for the government. The 1909 Wright A aircraft was very similar to the 1908 machine but it included a number of improvements. The water-cooled engine, practically unchanged, delivered about 30.6 horsepower. Each of the four cylinders, arranged vertically in line, had a 4-inch stroke and a 4 3/8-inch bore. The aircraft had modified controls, allowing the engine to be controlled by a foot-pedal placed on the footrest crossbar midway between the two seats. The pedal, connected to the magneto, served to advance or retard the spark.

The oil reservoir held .66 gallons of oil circulated by a small engine-driven pump. The vertical radiator, constructed of brass. stood five feet high and nine inches wide and had a capacity of 2 1/2 gallons. The gasoline tank held 13 gallons, enough for a flight of 3 1/2 hours. The gasoline pump, geared to the engine shaft, forced liquid fuel through a small pipe into an open tube where it mixed with air to form gasoline vapor. The engine started when someone held a piece of waste saturated with gasoline over an open tube into which air flowed while the pilot simultaneously cranked the propellers.

After the engine started, it was necessary to let it warm up before launching the aircraft. Once the engine was warm, the pilot advanced the spark with his foot to full power and then took off. The pilot maintained a constant speed until he was ready to land. The aircraft had no engine or flight instruments. A string was tied to the elevator in front of the pilot. By watching how the string trailed in the slipstream, the pilot knew whether he was slipping or skidding. Then he pulled the string over his head to release engine compression and stop the engine.

Orville and Lt. Lahm made the first official test flight on July 27, establishing a record of one hour and 12 minutes. On July 30, Orville and Lt. Foulois made a 10-mile cross-country flight, averaging 42.5 miles per hour, and winning a bonus of $5,000 above the $25,000 contract price. The Army accepted its first aircraft, Signal Corps Airplane Number 1, on August 2 1909.

Other pioneers were also flying during this period. Glenn H. Curtiss, born in Hammondsport, New York, was the first American after the Wrights to build and fly an airplane. Curtiss had successfully built engines for bicycles and he raced motorcycles of his own design and manufacture, outfitting them with lightweight, air-cooled engines that he built. In 1904, Curtiss won the contract to build an engine for the Army's first dirigible, initiating his aeronautical career.

Curtiss' interest in aircraft led him to form the Aerial Experiment Association with Dr. Alexander Graham Bell, Lt. Selfridge, and others in 1907. The association built and flew several kites, gliders, and aircraft. On June 20 1908, Curtiss flew the organization's biplane, June Bug, a distance of 1,266 feet. Curtiss used an air-cooled V-8 engine that had evolved from his motorcycle work; the engine reached almost 40 horsepower.

When the Aerial Experiment Association dissolved, Curtiss joined with A.M. Herring to form the Herring-Curtiss Company, the first aircraft factory in the U.S. That partnership lasted only a few months, and Curtiss soon set up his own company. Curtiss' success with his seaplane designs overcame the U.S. Navy's skepticism about aircraft. The organization of Wright interests in Britain, France, and Germany attracted the attention of American financiers like Cornelius Vanderbilt and August Belmont. With their backing, the Wright Company was incorporated in November 1909.[6] Dozens of other airplane companies sprang up, but the lack of either a commercial or military infrastructure caused many to fail. By 1911, only a handful of firms were manufacturing airplanes, although nearly fifty companies were producing parts and supplies. Among the top companies were Burgess, the Wrights, and Curtiss.[7] Even though aviation manufacturing did not become monopolistic, corporate changes and mergers did occur. Upset by Wilbur's death in 1912, Orville sold the Dayton factory and patents in 1914. The purchasing syndicate merged in 1916, in anticipation of war contracts, with the Glenn L. Martin Company as the Wright-Martin Company. The same syndicate controlled the Simplex Automobile Company, and Wright-Martin was one of the first to adopt the production techniques of the automobile industry. Wright-Martin prospered as an engine, rather than airframe, producer. Glenn L. Martin left the firm to establish his own company in

1917. By 1917, prospective war contracts led to consolidation of most of the smaller companies into major groups such as the Aeromarine Plane and Motor Company, Burgess, Curtiss, L.W.F. Engineering Company, Standard Aircraft Corporation, Sturtevant Aircraft Corporation, and Wright-Martin.[8]

PROLOGUE REFERENCES

[1]
Histories of aviation abound, including many excellent studies. The excellent ones invariably include useful bibliographies and references that provide additional sources for the interested reader to consult. Bibliographies of aviation history also exist, including Dominck A. Pisano and Cathleen S. Lewis, Air and Space History: An Annotated Bibliography (New York: Garland Publishing, Inc., 1988.)

For Air Force history, see Jacob Neufeld, Kenneth Schaffel, Anne E. Shermer, Guide to Air Force Historical Literature, 1943-1983 (Washington, DC: Office of Air Force History, 1985); and the earlier work by Mary Ann Cresswell and Carl Berger, United States Air Force History: An Annotated Bibliography (Washington, DC: Office of Air Force History, 1971.) The Air Force Historical Research Center at Montgomery, Alabama, is a treasure house of source materials on aviation history; it is accessible either directly or through any major Air Force command's history offices.

Another good starting point is with any of the writings of Richard P. Hallion, an American resource in aviation history. He has published extensively, has mastered the publications in the field, and knows many if not most of those active in aviation. Start with his The Literature of Aeronautics, Astronautics, and Air Power (Washington, DC: Office of Air Force History, 1984.) His published works explore many major aviation topics, including Test Pilots: The Frontiersmen of Flight (Garden City, NY: Doubleday & Co., Inc., 1981.)

The many scholarly works of Charles H. Gibbs-Smith are fundamental, including Aviation: An Historical Survey from Its Origins to the End of World War II (London: Her Majesty's Stationery Office, 1970); and The Invention of the Aeroplane, 1799-1909 (London: Her Majesty's Stationery Office, 1966.) Readable works are provided by Tom D. Crouch, including A Dream of Wings: Americans and the Airplane, 1875-1905 (Washington, DC: Smithsonian Institution Press, 1981); and The Bishop's Boys: A Life of Wilbur and Orville Wright (New York: W.W. Norton & Co., 1989.)

Roger E. Bilstein's Flight in America, 1900-1983: From the Wrights to the Astronauts (Baltimore: The Johns Hopkins University Press, 1984) is a readable, useful history by a scholar in the field. Scholarly essays providing an overview of aviation development are in Eugene M. Emme, ed., Two Hundred Years of Flight in America: A Bicentennial Survey (San Diego, CA: Univelt, Inc., 1977.)

Almost all aviation histories are illustrated with drawings and photographs that make it easier for the general reader to visualize the events and the technology under discussion. One such delightful publication, dated and internally contradictory in places, is by the editors of Year, Flight: A Pictorial History of Aviation (Los Angeles, CA: Year Incorp., 1953.) The Army Air Forces' Historical Office published an excellent pictorial history, The Official Pictorial History of the Army Air Forces (New York: Arno, 1979 -- a reprint of the 1947 book published by Duell, Sloan, and Pearce.) Many other pictorials, official and unofficial, preceded and followed these two examples.

[2]
A brief recitation of the Army's early aviation is provided in Alfred Goldberg's A History of the United States Air Force, 1907-1953 (Princeton, NJ: D. Van Nostrand Co., Inc., 1957.) Longer and more detailed information on the subject is in Juliette A. Hennessy's The United States Army Air Arm, April 1861 to April 1917 (Air University: Research Studies Institute, May 1958), published as USAF Historical Study Number 98. The book by Charles deForest Chandler and Frank P. Lahm, How Our Army Grew Wings: Airmen and Aircraft Before 1914 (New York: The Ronald Press Co., 1943) includes several useful appendices. Another "early bird" source is Benjamin D. Foulois, From the Wright Brotherws to the Astronauts: The Memoirs of Major General Benjamin D. Foulois (New York: McGraw-Hill Book Co., 1968.)

[3] For the Balzar-Manly engine. see Robert B. Meyer. Jr., ed., <u>Langley's Aero Engine of 1903</u> (Washington, D.C.: Smithsonian Institution Press. 1971.) See also, C. Fayette Taylor, <u>Aircraft Propulsion: A Review of the Evolution of Aircraft Piston Engines</u> (Washington, D.C.: Smithsonian Institution Press, 1971.) Both of these paperback books are in the Smithsonian's Annals of Flight.

[4] Works on the Wright brothers begin with their collected writings. A selection is in Fred C. Kelly's <u>The Miracle of Kitty Hawk: The Letters of Wilbur and Orville Wright</u> (New York: Farrar, Straus, and Young, 1951.) Kelly also wrote an authorized biography, <u>The Wright Brothers</u> (New York: Harcourt, Brace & Co., 1943.) See Crouch's work on the Wrights cited above, <u>The Bishop's Boys</u>. A work focused on the brothers' scientific and engineering prowess is Harry Combs, <u>Kill Devil Hill: Discovering the Secret of the Wright Brothers</u> (Boston: Houghton Mifflin Co., 1979.) Another useful work is Richard P. Hallion, ed., <u>The Wright Brothers: Heirs of Prometheus</u> (Washington, D.C.: Smithsonian Institution Press, 1978), a collection of insightful essays on the Wrights.

[5] For the Wrights' engines, see Leonard S. Hobbs, <u>The Wright Brothers' Engines and Their Design</u> (Washington, DC: Smithsonian Institution Press, 1971); and C. Fayette Taylor, <u>Aircraft Propulsion: A Review of the Evolution of Aircraft Piston Engines</u> (Washington, DC: Smithsonian Institution Press, 1971.) Both of these paperback books are in the Smithsonian's Annals of Flight publications. Assessment of the brothers' achievements was also provided by M.P. Baker, "Wright Brothers as Aeronautical Engineers," <u>SAE Quarterly Transactions</u>, V:1 (Jan. 1951), 1-17. This article included discussion by an Air Force Power Plant Laboratory expert, Opie Chenoweth, "Powerplants Built by Wright Brothers."

[6] By November 1910, construction was complete on the first of two factory buildings in West Dayton, and the Wright Company was turning out two airplanes a month.

[7] In June 1915, Grover C. Loening, aeronautical engineer with the Signal Corps, inspected and reported on the Wright, Curtiss, Burgess, and Thomas airplane plants. Most of these companies had foreign orders to fill. Curtiss plant's capacity far exceeded the others in output. Wright Company was capable of producing only one aircraft a week and Orville stated that he did not try to get war orders because he did not have an engine powerful enough for European demand. Loening also examined a number of new engines. After reporting to the Signal Corps, Loening resigned and joined the new Sturtevant Aeroplane Company in Massachusetts. Hennessy, 134.

[8] Bilstein. 29-30. Tom D. Crouch, <u>The Bishop's Boys: A Life of Wilbur and Orville Wright</u> (New York: W.W. Norton & Co., 1989), 465-467.

Glenn L. Martin's company merged with the Wright Company to form the Wright-Martin Aircraft Corporation in 1917. He withdrew from the partnership in 1918 to set up his own company, and later built Martin bombers for the U.S. government. Wright-Martin Company was reorganized again in 1919 as the Wright Aeronautical Company. In 1929, Wright Aeronautical and Curtiss Aeroplane and Motor merged to form Curtiss-Wright, the second largest U.S. manufacturer of aircraft and engines.

CHAPTER I

THE AIR ARMY, 1910 - 1917

Signal Corps Aviation School

As part of the government contract, Wilbur Wright trained two officers to operate the Army's first aircraft at College Park, Maryland, in October 1909.[1] The two pilots, Lieutenants Frank P. Lahm and Frederic E. Humphreys, crashed the machine in November. They subsequently returned to their regular assignments. However, before the crash, Lt. Benjamin D. Foulois returned from France in time to train. While the aircraft was in repair, the Army moved flying operations to Fort Sam Houston, Texas, for the winter. As the only Army officer on flying duty with the Aeronautical Division, Foulois and a party of enlisted men took the aircraft to Texas, arriving in February 1910.

Foulois began flying in Texas in March 1910. Since he had not yet soloed, he corresponded with the Wrights as problems developed. The Signal Corps allotted only $150 in 1910 for aviation gasoline, oil, and repairs, so Foulois covered some expenses with his own money. For several months in 1911, Robert F. Collier, the publisher, lent his Wright B aircraft to the Signal Corps for use at Fort Sam Houston. The Wrights sent an instructor to help Foulois master the newer model. Collier's loan proved timely because the original Army flyer transferred to the Smithsonian for exhibition.[2]

Beginning in 1908, the Signal Corps requested $200,000 each year for aviation, but Congress did not appropriate any money specifically for aviation until 1911. The Army had only one pilot and one aircraft up to that point in time. The money used to purchase the Wright airplane came from an experimental fund. The Signal Corps took money from a general fund for maintaining military telephone and telegraph installations to support flying. Finally, in March 1911, Congress appropriated $125,000 for 1912, making $25,000 immediately available.[3]

The Chief Signal Officer used the money to acquire five aircraft, all pusher-types.[4] Three of these were Wright B models, one of which was built by W. Starling Burgess. The other two were Curtiss planes. The Curtiss aircraft were equipped with a Curtiss 51.2 horsepower, 8-cylinder, water-cooled engine with a single pusher propeller located behind the center section of the wings. Both the Curtiss and Wright companies sent pilots to instruct the Army flyers.[5]

In May 1911, after the first flying fatality at Fort Sam Houston, the commanding general prohibited further flying from the drill ground. Meanwhile, the Signal Corps began constructing a flying school at College Park. The men and airplanes moved back to Maryland in June and July. Needing a test to qualify pilots, the Army adopted the regulation of the Federation Aeronautique Internationale as administered by the Aero Club of America. At the end of 1911, the Signal Corps aviation fleet consisted of five airplanes, two balloons, and a dirigible. The personnel strength of 23 persons included six pilots.[6]

The Army recognized that it needed improved aircraft if the air arm was to operate in the field. In drawing up new specifications, the Army found that the two most desired characteristics, greater weight-carrying capacity and higher speed, could not coexist in a single aircraft due to limitations of available power plants. Therefore, the Chief Signal Officer therefore issued specifications for two types of aircraft, a "Speed Scout" for strategic reconnaissance and a "Scout" for tactical reconnaissance, in February 1912. Considering these requirements, the Signal Corps ordered Scouts from Curtiss, the Wright Company, and the Burgess Company (the first tractor aircraft). It also ordered two Speed Scouts from the Wright Company. The Wright C Scout, slightly larger than the Wright B model, was stronger and powered by a 6-cylinder, 50-horsepower Wright engine.[7]

The pilots at College Park advanced their knowledge of flying and aircraft throughout 1911 and 1912. The pilots' first accomplished cross-country and altitude flights, conducted aerial photography experiments, and tested a bombsight. Later they experimented in night flying and in firing a machine gun from an airplane.[8] Aircraft maneuver tests with ground troops also occurred. The number of lieutenants placed on aviation duty increased from the funds available in 1912. A new training policy assigned them were sent to airplane manufacturing plants for shop and flying courses.

By November 1912, the College Park school had nine aircraft, including "hydroplanes," that is, land planes equipped with pontoons so that they could operate from both land and water. One such seaplane, built by the Burgess Company, was the Army's first tractor model. Atractor model featured a propeller mounted in front rather than behind the wings. That winter, the Wright airplanes and pilots relocated for the second time to Augusta, Georgia. However, at the invitation of Glenn Curtiss (who moved his flying activities to California in 1910 to take advantage of the better weather there), the Curtiss planes and pilots went to San Diego. Curtiss established a flying school on North Island in San Diego Bay in 1911, and invited the Army and Navy to send officers for free instruction as pilots. In November 1912, the Signal Corps set up a school on North Island and closed College Park. The Signal Corps officially designated North Island (later called Rockwell Field and used by the Army until October 1935) as its aviation school in December 1913, the Army's first permanent aviation school. Progress in military aviation thereafter became more focused.[9]

Meanwhile, strained relations with Mexico over the revolutionary regime of General Victoriano Huerta, resulted in the five flyers at Augusta, Georgia, being ordered to support the 2nd Division. Captain Charles deForest Chandler assembled the 1st Aero Squadron in Texas City, Texas, the first air combat unit in the U.S. Army. Although the pilots did considerable flying, operations ceased in November 1913 without the squadron's engaging in battle. After the crisis with Mexico subsided, the 1st Aero Squadron moved to San Diego where it joined the rest of the Army's air strength at the aviation school. In addition to flying training, the school began to emphasize ground training. Special activities included further experiments with bombsight and parachutes, and altitude and cross-country flights. Mechanics worked at the school on improving engines, including experiments with higher test gasoline and different lubricating oils. Signal Corps specifications called for engines to pass a six-hour test at the Bureau of Standards.[10]

For the aviation school, 1914 was a difficult year.[11] Accidents at San Diego caused the Inspector General to review the Signal Corps Aviation School and the 1st Aero Squadron for the first time in February 1914. Both units received unfavorable reports. The IG was critical of the aircraft manufacturers, especially the Wright Company, and recommended Glenn Martin as a progressive builder of aircraft. The IG further recommended that the government establish a factory of its own with Martin in charge because American manufacturers were lagging.

The Wright and Curtiss companies made efforts in 1914 to improve their machines. The Wright Company substituted a steering wheel on its new tractor-type aircraft for the lever previously used for wing warping. They equipped it with a 120-horsepower Austro-Daimler engine controlled by levers located on a sector of the steering wheel. The Curtiss Company produced an 8-cylinder, 160-horsepower engine, one of the highest powered engines ever produced by an American manufacturer.[12]

In 1914, the Signal Corps condemned the pusher-type airplanes involved in most of the twelve fatalities that occured out of forty-eight officers detailed to flying. Pushers had a record of diving and crashing. On those occasions, the engine tended to tear loose from the rear of the plane, ramming the pilot and passenger. The action left the Army with only five trainers at San Diego and even those machines needed reconditioning. Training continued, however, with the help of a sport plane acquired from Martin converted into a dual-control trainer. Also at this time, the Army ordered the JN-1 from Curtiss, first of a series of trainers known familiarly to a generation of student pilots as the "Jenny."[13]

The school adopted a policy of employing civilians to give preliminary instruction in flying. The results proved satisfactory. In addition, the school taught the care and repair of aircraft and engines. Several experts lectured on meteorology, aeronautical engineering, propellers, and internal combustion engines.

On July 18, 1914, legislation gave statutory recognition to Army aviation by creating an Aviation Section, as part of the Signal Corps. The act authorized a strength of 60 officers and 260 enlisted men. The section was responsible for operating and supervising all military aircraft, including balloons,[30] airplanes, and necessary equipment. It was also responsible for training officers and enlisted men. The act limited officers to unmarried lieutenants, incorporated earlier provisions for flying pay, and established the aeronautical ratings of junior military aviator and military aviator.

Also in July, Grover C. Loening, first in the U.S. to receive a master's degree in aeronautics (Columbia University, 1910), was appointed aeronautical engineer in the Signal Service at large. He resigned from the Wright organization to set up the Army's first aeronautical engineering program at the San Diego school. This appointment, he wrote, was[14]

> *the beginnings, by me, of the establishment of an engineering division which in a few short years was to grow into the organization at McCook Field now known as the Air Corps Engineering Division, subsequently moved to Wright Field, Dayton, Ohio.*

Loening remained for a year before returning to private industry with the Sturtevant Aeroplane Company. He had a free hand at reorganizing the system of training, engineering, maintenance, and procurement. In addition, Loeining established an Experimental and Repair Department responsible for all experimental work and development of new airplane construction features. The Training Department included both airplane and engine repair sections and a school section. The school section, which gave theoretical and practical training in aircraft and engines, came under George E.A. Hallett, an engine expert who had previously worked for Curtiss. Although maintenance functions absorbed a significant part of the staff's time, development of aircraft and engines also played a large role. This effort was the first injection of sound engineering into the Signal Corps' aviation program, replacing the trial-and-error approach of aircraft design and maintenance. It marked the start of developing a professional engineering methodoloy for maintenance and for aircraft research and development.[15]

Because of difficulties with its assorted early aircraft, the Army tried to acquire a standard airplane to improve safety and performance.[16] In 1914, the Chief Signal Officer received authorization to hold a competition for three aircraft to be purchased at $12,000, $10,000, and $8,000 depending on how well each one satisfied the requirements for the reconnaissance aircraft. Requirements specified a two-place tractor biplane capable of lifting a load of 450 pounds and having a high speed of 70 miles per hour. Other requirements included a 4,000-foot climb in 10 minutes while fully loaded and sufficient fuel and oil for four hours of nonstop flying. Twelve companies made offers but only two, Curtiss and Martin, delivered. Only Curtiss Aeroplane Company's entry fulfilled every requirement, and the Chief Signal Officer cancelled the competition.[17]

Airframes and engines were developing too fast to standardize, and manufacturers were too small to build aircraft in quantity. Nonetheless, establishment of specifications was a significant step forward in recognizing the value of the aircraft as a military weapon. Equally important was the development of a procedure to select the weapon: a competition between airplanes submitted by manufacturers in which they earned points for performance demonstrated in flight. The aircraft tests included measurements of speed, rate of climb, maneuverability, and field of vision. A separate board evaluated construction and standards of workmanship. Thus, an administrative mechanism for selecting weapons had evolved.[18]

By 1915, primary responsibility for the evolution of military aircraft rested with the small group stationed at North Island aviation center in San Diego Bay. The center had two departments, one for training, the other for experiment and repair. The experimental department had a staff of eight people: one officer in charge, a civilian aeronautical engineer, a civilian mechanical engineer, and five civilian mechanics. This small organization provided the nucleus for expansion. Their duties at the center, in addition to the study of new types, included overhauling, repairing, and rebuilding the training aircraft as well as maintaining equipment for ground servicing. Meager appropriations continued to restrict the number of aircraft acquired and the number of pilots available to fly them.

In the 1914 competition, the lack of a reliable aircraft engine set limits on the performance of the new aircraft. The problem was critical enough for the Signal Corps to conduct an engine competition in 1915 similar to the aircraft contest. During these competitions and subsequent flight trials, the aircraft manufacturers began the practice of locating representatives at the aviation center to stay in touch with the needs of the air arm. Thus, by the end of 1916, the Signal Corps had a way of selecting weapons and a close working relationship with the manufacturers supplying them.[19]

After additional facilities had been completed for the San Diego school in 1915, technical instruction, limited earlier to engine overhaul and adjustment, was expanded. The school introduced a systematic course of instruction, added aircraft for teaching, and increased the number of civilians to provide continuous instruction for both officers and enlisted. In July, a board of officers at the school adopted requirements for the aviation mechanician examination, dividing them into two parts: airframe maintenance and repair; and engine construction, maintenance, and repair. Engine requirements demanded they learn how to clean an engine, grind its valves, adjust clearances, time the valves and spark, clean the magnetos, locate trouble in and repair the ignition system, adjust the carburetor, and locate and correct other types of problems.[20]

Maintenance functions absorbed much of the staff's time, as well as the development of new equipment. The aviation school at San Diego tested all new instruments and accessories for aircraft and demonstrated. It was the only aviation station where sufficient personnel and material were available for this purpose. Efficiency increased, evidenced by the fact that almost all pusher aircraft were gone and new flying records occurred frequently. Furthermore, rivalry between the Curtiss and Wright camps ended when the Signal Corps chose the Curtiss machines over the Wright Pushers for service.[21]

Little happened at the school early in 1916 other than routine training because of reorganization and the need to furnish personnel for overseas squadrons. Training in night flying began, and major developments with aircraft radio occurred. Students that completed training and received ratings went either to the 1st Aero Squadron or became permanent staff at the school. Then in April 1917, the Chief Signal Officer ruled that flying schools were to concentrate on training only. He discontinued any projects involving experimentation or non-flying activities.[22]

By the time the U.S. entered World War I, the size and activities of the Signal Corps Aviation School had increased considerably as had the number of planes and the students in training. The National Defense Act of June 1916 increased the strength of the Aviation Section and established a reserve corps of

14

officers and enlisted men. Flying enthusiasts in many states organized air units, assisted by the Aero Club of America. These units provided their own planes, often at their own expense. The only one officially recognized by the War Department as a National Guard unit entered into federal service from New York in 1916, but it remained active for only a few months. In 1917, these air units and reserve corps provided a manpower pool for the Aviation Section.[23]

Aviation Section[24]

American military aviation was still in its formative stages when Europe plunged into war on July 28, 1914. Lack of funding and support coupled with inferior equipment and difficult manpower problems caused Army aviation to lag compared to other nations. Safe from attack behind ocean barriers east and west, and faced by weak nations north and south, the United States did not share the intense spirit of international competition that characterized the military establishments in Europe before the war. By 1914, the U.S. had lost leadership in aeronautics, especially in military aviation.[25]

In 1911, five nations had more pilots than the U.S. France, for example, had 353 certified pilots compared to 26 in the U.S., only 8 of whom were military. Between 1908 and 1913, the world spent an estimated $85 million for military and naval aviation. Germany spent $28 million; France $22 million; Russia $12 million; Italy $8 million; Austria, $5 million; Great Britain $3 million; Belgium $2 million; Japan $1.5 million; and Brazil $500,000. During that period, the U.S. ranked fourteenth, spending $435,000. War Department appropriation for aeronautics in 1914 totaled only $125,000. The Army received about 39 airplanes by the end of 1914, but during the year, not more than 20 were serviceable.[26]

At the outbreak of the war, both sides had several times the number of military aircraft available to the U.S. (though sources differ on the exact numbers). European nations hurriedly built more aircraft to aid armies in battle, and placed orders with American manufacturers. President Wilson issued a proclamation in August 1914 forbidding various acts within the territory and jurisdiction of the U.S in an attempt to keep it neutral. The proclamation permitted production for belligerents. Thus, American aircraft and engine companies were able to supply the Allied nations and exclude Germany. U.S. industry benefited from increased production and gained experience, and some expansion in American airplane factories occurred. The results were not uniformly beneficial for the U.S, however, due to difficulty in acquiring planes from manufacturers working on foreign orders.[27]

The U.S. also had a problem keeping pilots due to the dangerous nature of flying itself. An underlying obstacle with Army aviation was the lack of a clearly defined status and function. Aeronautics was costly in terms of manpower, materiel, and facilities, and the Signal Corps increasingly found it difficult to support the air arm. Many flyers believed that Army aviation would never make satisfactory progress until it received statutory recognition. A bill introduced into the House of Representatives in 1913 called for establishing aviation as a separate arm of the Army, but the War Department opposed the bill. Among pilots, only one favored immediate separation from the Signal Corps; the others felt that aviation had not yet sufficiently matured.[28]

Six months after the Aviation Section formed, personnel strength amounted to 30 officers (2 non-flying Signal Corps officers, 18 junior military aviators, and 10 aviation students), 119 enlisted, and 8 civilian employees. The section had 11 aircraft; three assigned to the school and eight to serve with the 1st Aero Squadron. In addition, a school machine was under construction and an experimental machine was en route to San Diego. An order issued that month to equip the 1st Aero Squadron with eight new machines would raise the total to 21 aircraft.[29]

The Air Arm in Mexico

The 1st Aero Squadron, which had formed provisionally early in 1913, reorganized officially at San Diego in September 1914 in accordance with a War Department order. Under Captain Foulois, the squadron engaged in training and testing activities during 1914[30] and 1915. Prior to U.S. entry into World War I, this squadron represented the Army's tactical air strength. A few months after arriving at Fort Sam Houston in November 1915, the 1st Aero Squadron became the first and only Army tactical air unit before World War I tested under full field conditions.[31]

Pancho Villa's emergence as a revolutionary leader in Mexico once more made the Texas border an area of tension and potential conflict. On 9 March 1916, he raided Columbus, New Mexico, killing 17 Americans. The U.S. government directed Brigadier General John J. Pershing to pursue Villa into Mexico and take him dead or alive. Ordered to Columbus, the 1st Aero Squadron arrived there on 15 March. Foulois had 10 pilots, 84 enlisted, and 8 Curtiss JN-3 observation aircraft with 90-horsepower engines. By May, unit strength reached 16 officers and 122 men.[32]

The squadrons aircraft could hardly function in the mountains of northern Mexico, and by the end of March, the aircraft had demonstrated their inability to perform. Underpowered engines limited climbing ability with a military load, making it unsafe to operate in the mountains. The planes could not fly across the 10,000 to 12,000-foot mountains, and they could not operate in the high winds and storms. The dry atmosphere caused the wooden propellers to warp and come apart. Aerial scouting occasionally proved useful to Pershing's troop movements. Generally, the squadron could make short flights in good weather with mail and dispatches. The operation demonstrated the lack of experience and preparation of American military aviation.[33]

By late April, only two of the eight planes were still operational and Foulois burned both to avoid flying them any more. The replacement planes proved no more capable. As the new Curtiss N8 planes (duplicates of JN-4s being made for overseas use) arrived, the squadron used them in experimental and testing activities, but Foulois found the new aircraft unsuitable for field service. Even the 12 Curtiss R-2 aircraft with 160-horsepower engines proved unsatisfactory due to difficulties with the propellers and engines that required modifications on every plane. Curtiss twin-engine aircraft and Standard H2s were not much better operationally. Operations over Mexico lessened after the summer of 1916, but the squadron continued at Columbus until departing for Europe in August 1917.[34]

The fiasco of the 1st Aero Squadron's operations in Mexico, at a time when aviation in Europe was demonstrating great combat potential, dramatized the appalling shortcomings of Army aviation. The Aviation Section only had a reconnaissance plane when the U.S. entered WWI in April 1917. Although it was experimenting with other types of planes, it was clear the U.S. was far behind European countries in aeronautics. In addition, few aircraft engines were available in the U.S. and the war prevented purchasing foreign-made engines. The military finally recognized a need for more aircraft and properly trained pilots. In August 1916, Congress provided over $13.8 million for military aeronautics, a huge sum for the time. The funds included $600,000 for purchasing aviation sites.[35]

The Army desperately needed the money to train more flyers and to buy better aircraft. Unfortunately, improved planes were not available because the American aircraft industry had also fallen far behind in development and production of aircraft engines, the key to improved performance. The best engines had come from Europe even before 1914. By 1916, the warring nations needed more engines than they could produce. Planes built in America for the Allies usually had European engines installed in them only after their arrival abroad.[36]

Organizing for WWI

Preparing for war was difficult. The Army lacked important information for planning an aviation program. It did not send trained observers to Europe to gather available information about the technical and operational aspects of the war. Much of the data would have been inaccessible, however, because the Allies strictly censored aviation information and refused to permit American air observers to visit the front. Even if observers were at the front, American military observers did not possess enough experience in aviation to make reports of value. Air officers in Washington had never even seen a fighting plane. Reports on operations, organization, or equipment of the flying corps abroad were limited. The Aviation Section had to rely almost entirely on current periodicals and other literature for information. These sources, however, contained little technical information. The National Academy of Sciences pressed the government to form a national aeronautical laboratory as several European nations had done. The disparity in aeronautics and the onset of war in Europe caused Congress to append a rider to the Naval Appropriations Act. President Wilson signed the act on March 3, 1915, authorizing him to form a National Advisory Committee for Aeronautics (NACA).

NACA's mission was to supervise and direct the scientific study of flight in order to achieve practicable solutions and to direct and conduct research and experiment in aeronautics. The committee's 12 members included five people from outside government and representatives from the Army, Navy, Smithsonian, Weather Bureau, and the Bureau of Standards. Initially, NACA lacked research laboratories, making its primary role advisory.[37]

Following passage of the National Defense Act of June 1916, several new civilian agencies similar to NACA arose. The National Academy of Science formed a National Research Council in September to further scientific research applicable to both war and peace. The NRC conducted some research in aeronautics, but when wartime research became too extensive, the NRC recommended that the Signal Corps establish a Science and Research Division in February 1917.[38]

A Council of National Defense, consisting of the Secretaries of War, Navy, Agriculture, Interior, Commerce, and Labor, formed in August 1916. President Wilson reorganized this council in March 1917 to include an advisory commission of seven civilians with special knowledge of industrial and commercial resources of the country to apply itself to industrial and economic mobilization.[39]

Through the efforts of NACA and the Council of National Defense, a Manufacturers Aircraft Association formed in February 1917. The association worked to arrive at cross-licensing agreements satisfactory to all holders of patents involving aircraft material. Such an agreement took place in March 1917 and took effect the following July. The government recognized aileron patents held by the Wright-Martin and Curtiss companies, and offered each company $2 million for them (later reduced to a total of $2 million). Members of the Manufacturers Aircraft Association were to pay the association $200 for each aircraft built until the two companies received payment. This allowed Wright and Curtiss to finally end a battle over patent rights that had plaqued the aviation industry for quite some time.[40]

Three days before the U.S. declaration of war, NACA surveyed the aircraft industry in the U.S. The telegram sent to every aircraft manufacturer revealed the lack of organization and policy. The Secretary of the Navy, Josephus Daniels, after negotiating with Dr. Charles D. Walcott, chairman of the NACA executive committee, and with the concurrence of the Secretary of War, took the initiative in establishing a Joint Army-Navy Technical Board. The new board's responsibilities included standardizing the designs and general specifications of aircraft procured by the services.[41]

In 1916, the Chief Signal Officer recognized the existence of three different types of military aircraft to perform reconnaissance and direct artillery fire; combat; and pursuit.[42] Although the characteristics that differentiated the aircraft were vague, the recognition that aircraft had different roles was a step toward fuller development of the aerial weapon. The Chief Signal Officer believed that the most important function of aircraft was observation though he did concede that aircraft also had application for liaison and for defense against other aircraft. Accordingly, the squadrons he proposed to establish favored observation aircraft over bombers and other types.[43]

When the U.S. entered World War I, the nation lacked a knowledge of the mission expected of the aerial weapon and, thus, a knowledge of the specific types of aircraft needed. The limitations resulting from the shortage of funds and personnel had restricted growth to the point that little operating experience had been gained. The organization charged with developing the weapon was inadequate to the task. The aircraft and engine competitions providing for procurement on the basis of performance set important precedents, but formalization of the process did not occur.[44]

The U.S. needed a self-sufficient aircraft industry capable of exploiting and converting scientific and technical advances into designs for production. That is, the U.S. needed agencies for Research & Development, but in April 1917 an integrated system for aeronautical R&D was only beginning to emerge.[45]

Early in 1917, the Aviation Section of the Signal Corps only had about a half-dozen aeronautical engineers, several of whom were studying aeronautics at the Massachusetts Institute of Technology.[46] Only a dozen or so aircraft manufacturers existed in the U.S., and only about half of them had ever built as many as 10 aircraft. None had ever designed a successful combat aircraft. Curtiss Aeroplane and Motor Corporation, having constructed a large number of training aircraft for the War Department and for the British (supported by British engineers on the corporation's staff), was the only manufacturer with extensive experience. An effective aerial force could not function with such limited resources. Even if the War Department could have provided the manufacturers with definite performance requirements, the half-dozen designers could not have developed the types of aircraft required on the front. The immediate need for aircraft forced the U.S. to turn to Europe for models of tactical aircraft to put into production at once.[47]

Until 1916, the Signal Corps school at San Diego had largely served the Aviation Section's needs for flying training. The corps added new training facilities and new aero squadrons by April 1917. With increased appropriations and with war threatening, the need for new flying schools increased dramatically.[48] In addition, civilian flying schools began training reservists for the Aviation Section in late 1916. That year, the War Department authorized the organization of seven aero squadrons of 12 planes each, four in the U.S. and three overseas.[49] All of the squadrons were in existence early in 1917, but only the 1st was fully organized and equipped, and even its equipment left much to desire. Plans for expanding the air arm to 20 squadrons were still largely on paper at the outbreak of WWI.

Resources when the U.S. declared war on Germany were sparse. The Aviation Section had 131 officers, virtually all pilots and student pilots, and 1,087 enlisted men. It had fewer than 250 planes, all trainers by European standards, and five balloons. The Army did not have any bombers, fighters, or any of the combat types operating in Europe.[50]

The U.S. had never trained an aviator for actual combat overseas. No one who knew what kind of instruction was necessary for radio operators, photographers, or regular enlisted. Consequently, the first men charged with the training program had to learn by trial and error before teaching others. Distribution of tactical units among seven aero squadrons occurred, but most were not fully organized or equipped. Neither balloon companies nor National Guard units existed, and only two reserve aero squadrons were available.[51]

Serious difficulties lay ahead.

CHAPTER 1 REFERENCES

Information on the air Army in this period is based, for the most part, on the detailed work of Juliette A. Hennessy, The United States Army Air Arm, April 1861 to April 1917 (Air University: Research Studies Institute, May 1958), published as USAF Historical Study Number 98. Much of this material is also covered by Charles deForest Chandler and Frank P. Lahm, How Our Army Grew Wings: Airmen and Aircraft Before 1914 (New York: The Ronald Press Co., 1943), a book with several useful appendices. Benjamin D. Foulois, From the Wright Brothers to the Astronauts: The Memoirs of Major General Benjamin D. Foulois (New York: McGraw-Hill Book Co., 1968), another personal account of an "early bird," provides a countervailing view of Billy Mitchell.

A succinct overview of this period is provided by Alfred Goldberg's A History of the United States Air Force, 1907-1953 (Princeton, NJ: D. Van Nostrand Co., Inc., 1957). For an account of the Army and Navy at Rockwell Field, see the pictorial by Elretta Sudsbury, et al., Jackrabbits to Jets: The History of North Island, San Diego, California (San Diego, CA: Neyenesch Printers, Inc., 1967. Informative discussion of aviation engines used by the Army in the pre-WWI period, and Hallett's role at North Island, was provided by Robert B. Casari, "Powering Army Aircraft Through 1915," Journal American Aviation Historical Society (Summer 1977), 91-103.

For discussion of how the absence of a meaningful doctrine arrested development of military aircraft before and after World War I, see I.B. Holley, Jr., Ideas and Weapons (Washington, DC: Office of Air Force History, 1983), a book originally published by Yale University Press in 1953. Broader application of this thesis on the critical role of doctrine in the evolution of the Air Force is in Robert F. Futrell's Ideas, Concepts, Doctrine: A History of Basic Thinking in the United States Air Force, 1907-1964 (New York: Arno Press, 1980). A wider view of aviation developments in this period is provided by Roger E. Bilstein, Flight in America, 1900-1983: From the Wrights to the Astronauts (Baltimore: The Johns Hopkins University Press, 1984).

[1]   During 1910, the Wrights organized flying schools at Huffman Prairie and Montgomery, Alabama, the latter for the winter months. The training field at Huffman Prairie was largely cow pasture with a woooden shed at one end of the field that served as a hangar.

Flying machines cost between $5,000 and $7,500, could carry only a small load, and lacked permanent landing fields, hangars, and maintenance facilities. The absence of infrastructure led the Wrights and Curtiss to organize their own exhibition companies and hire pilots to fly for them. The Wright Exhibition Company, playing carnivals, circuses, country fairs, and other events where crowds could be expected, grossed a million dollars. Though the Wrights withdrew from flying exhibitions in 1911 after two seasons, other exhibition teams, like the one organized by Curtiss, continued to fly. Soon specially organized flying meets became major spectator events, attracting thousands of people in a single day. Bilstein, 16-18, 26.

[2]   Hennessy, 39-40. Goldberg, 4-6. Foulois, 1-5, 68-85.

[3]   At the same time, the Navy received its first appropriations for aeronautics and used the $25,000 to purchase its first three aircraft. Hennessy, 40, 42, 108. Goldberg, 6.

[4]   The Army purchased a total of 15 airplanes from the first appropriation. Hennessy, 108.

[5]   Hennessy, 42, 45. Foulois, 85-86, 91-94.

The Curtiss aircraft, unlike the Wright aircraft, was a one-seater, forcing students to learn to fly alone. After mastering the controls, students taxied around until they were experienced enough to takeoff.

The Curtiss throttle operated like the accelerator pedal on automobiles. Pressing down with the foot, speeded the engine; relaxing pressure slowed the engine. To make flying easier for the pilot, an officer in 1911 supplemented the foot control on the Curtiss plane with a hand throttle similar to that on the Model T Ford; it could be used to set the engine speed at any desired rate.

The Wright foot throttle worked the opposite way. The engine was throttled down by pushing with the foot. The engine had little compression and, when the pilot glided in for a landing with the engine throttled, it continued to pump gas, which spilled over the side of the engine. Because the gas then ran down on the wing, a metal pan was placed beneath to catch it. When the Wright plane landed, the pilot gave it more gas in order to taxi in. Since dripping gas often caught fire, ground crews had to stand by with fire fighting equipment. Hennessy, 50, 53. Goldberg, 6. Bilstein, 21.

[6]   Hennessy, 45. Goldberg, 6. Foulois, 94.

[7]   Hennessy, 58. Frey, 23. Foulois, 97.

The "Speed Scout" was to be a fast single-seat aircraft used for "strategic" reconnaissance, that is, to locate and report on troop movements beyond the immediate vicinity of friendly forces. It was to have a radius of operation of about 100 miles, a minimum speed of 65 miles per hour, and enough fuel for one hour.

The second type, the "Scout," was to be used for "tactical" reconnaissance, that is, for reporting on enemy forces that were approaching or in contact. Minimum speed was to be 45 miles per hour. Rate of climb was to be 2,000 feet in 10 minutes with a load of 450 pounds. With a load of

600 pounds, speed could be reduced to 38 miles per hour and the rate of climb lowered to 1,600 feet in 10 minutes.

Both aircraft had to be easily transportable as well as rapidly and easily assembled. Their engines had to undergo a two-hour flight test and be capable of being throttled -- that was in contrast to previous practice of having only two speeds, idle and full power. Both had to be able to land and takeoff from plowed fields and to glide to safe landings from 1,000 feet with a dead engine. The starting and landing device for each was to be part of the machine.

[8]   Goldberg, 7. Foulois, 95-97, 99-100.

Despite successful flight testing of the Lewis machine gun, standard on Allied aircraft in World War I, none could be acquired by the Signal Corps at this time because the Army had not yet adopted it. In 1912, the Signal Corps began buying some foreign-made engines such as the 103-horsepower Renault, the 120-horsepower Austro-Daimler, and the 160-horsepower Gnome. The Bureau of Standards did not have a dynamometer that could handle over 100 horsepower. Lt. Arnold, engineering officer at College Park, made his own, connecting an electric dynamometer to a water dynamometer in order to check out the engine.

To determine engine revolutions-per-minute in this early period, the engine was run up to maximum speed at full power, and a revolution counter on the propeller shaft and a stop watch were used to determine the number. When a more accurate method of measuring power was needed, the aircraft was tied to a rope fastened to a spring scale and the scale was attached to a stake in the ground. The engine was speeded to full power and propeller thrust was indicated in pounds on the scale. A tachometer satisfactory for installation in an aircraft was developed by August 1915. Frey, 24, 55.

[9]   Hennessy, 86. Goldberg, 6-8. Sudsbury, 5-29, 32, 35, 190-191.

[10]   Frey, 40, 46. Foulois, 107-110.

Two overseas training schools had been established, one in the Philippines and one in Hawaii. Lahm, who transferred to the Philippines at the end of 1911, opened a flying school at Fort William McKinley near Manila in March 1912. Using a Wright B airplane, Lahm trained several pilots during the next two years. The school at Fort Kamehameha in Hawaii, established in the summer of 1913, was less successful because of trouble with the seaplanes used for training, lack of facilities, and the winds in the area. The latter school lasted for little more than a year. Hennessy, 79-85. Goldberg, 7-8.

[11]   Hennessy, 103.

Rivalry flourished between pilots flying the various types of planes at the school, especially the Wright and Curtiss aircraft. In August 1914, the school prohibited both the military and civilian employees from publicly discussing the relative merits of aircraft and engines from different companies. Hennessy, 102, 127.

[12]   Hennessy, 104.

[13]   Hennessy, 103. Goldberg, 8-9. Foulois, 114. Loening, Our Wings, 50-51.

[14]   Grover Loening, Our Wings Grow Faster (Garden City, New York: Doubleday, Doran & Co., Inc., 1935), 48, 51.

[15] Hennessy, 122, 137, 139, 141. Frey, 44, 46, 51-52, 66. Edward O. Purtee, History of the Army Air Service, 1907-1926 (WPAFB, OH: Air Materiel Command, 1948), 15. Grover C. Loening, Our Wings Grow Faster (Garden City, NY: Doubleday, Doran & Co., Inc., 1935), 48-50, 63.

Captain V.E. (Virginius Evans) Clark enrolled in the air engineering course at MIT at his own expense, returning in July 1915 to the San Diego school as the first aviation officer to graduate from the institution with a degree in aeronautical engineering. He was assigned to duty in the Experimental and Repair Department. Frey, 54. Hennessy, 124.

Curtiss rented a motor boat for use at North Island, and the boat came with Hallett, an engine mechanic. Hallett, who retired from the Army a few years after World War I, was hired by Curtiss in 1911 as an engine mechanic. Thereafter, Hallett naturally assisted Curtiss in working on aircraft at the island. Sudsbury, 13, 29, 32, 34.

Discussion of aviation engines used by the Army in the pre-WWI period, and Hallett's role at North Island, was provided by Robert B. Casari, "Powering Army Aircraft Through 1915," Journal American Aviation Historical Society (Summer 1977), 91-103.

[16] Holley argued that the emphasis on a standard aircraft was clear evidence that in 1914 the Army lacked an understanding of aerial doctrine, namely the concept that tactical aircraft types needed to be markedly different in order to perform different missions.

[17] Hennessy, 117. Holley, 33-34. Foulois, 115-116.

[18] Holley, 34.

[19] Holley, 33-35. Frey, 54.

[20] Hennessy, 138, 141, 144. Frey, 51-52, 54.

[21] Hennessy, 137-138.

[22] Hennessy, 158, 166. Frey, 79.

Flying safety greatly improved at North Island. From January 1, 1915 through February 1916, 1,682 hours of flight were logged in seven aircraft without a fatality. From January 1 to December 26, 1916, 6,087 flights were made totaling 3,356 hours and 251,775 miles without a fatality. Frey, 67, 71. Hennessy, 166.

[23] Hennessy, 133-134, 158, 183, 185-187. Goldberg, 11.

[24] The charts of the period between 1915 and 1917 show the Aeronautical Division rather than the Aviation Section as the Signal Corps' official organization for air. Purtee, 20.

[25] Goldberg, 8.

[26] Hennessy, 112. Foulois, 111. Martin P. Claussen, Materiel Research and Development in the Army Air Arm, 1914-1945, Army Air Forces Historical Study Nr. 50 (Hq. Army Air Forces: Historical Office, Nov. 1946), 12.

[27] Hennessy, 125, Bilstein, 31-32.

[28] Hennessy, 109. Goldberg, 8. Frey, 47. Foulois, 103-105, 111. The one dissent was by Captain Paul W. Beck who urged a separate organization for aviation. Interestingly, Captain William Mitchell, who later commanded air operations at the front in France during World War I and after the war was court-martialed for his supporting an independent air force, was among the officers, including Foulois, who protested against removing control of aviation from the Signal Corps.

[29] Hennessy, 124.

[30] Meanwhile, in 1914 relations with Mexico had once again reached a critical stage, providing the occasion for the first American use of aircraft in actual military operations by the Navy. In April 1914, Curtiss flying boats from the battleship Mississippi and the cruiser Birmingham operated in the Gulf. In April 1914, the 1st Aero Squadron sent a detachment of five officers and three aircraft to Galveston, Texas, to join the U.S. expedition against Vera Cruz, but it arrived too late to catch the transport and the aircraft were never unpacked. The detachment returned to San Diego in July 1914. Bilstein, 32-33. Hennessy, 105-106. Goldberg, 9.

[31] Goldberg, 10.

[32] Hennessy, 156. McFarland, viii. Frey, 73. Foulois, 122-126.

[33] Hennessy, 169. Bilstein, 33. Goldberg, 10. Frey, 74-75. Foulois, 126-129.

[34] Hennessy, 172-176. Frey, 73-74. Foulois, 133-136.

[35] Hennessy, 154. Goldberg, 10. Foulois, 136-137.

[36] Goldberg, 11.

[37] Hennessy, 130-131. Holley, 66. Bilstein, 30-31.

[38] Hennessy, 155. Holley, 112-113.

[39] Hennessy, 155-156.

[40] Hennessy, 156. Bilstein, 28. McFarland, 569-571. The Manufacturers Aircraft Association is often referred to as the Aircraft Manufacturers Association in the literature.

Following payment of both companies, fees of not more than $25 each were to be paid on all aircraft. Other subscriber owners of designs or devices used by another manufacturer would be paid one percent of the contract price until the owner received a total of $50,000. By the cross-license agreement, subscribers granted each other licenses under all airplane patents except foreign and cerain specified patents. The association was designated as the agent of the subscribers to execute licenses accordingly, and each member agreed not to enter into any arrangement that would restrict their agreements.

[41] Holley, 39-40.

[42] Hennessy, 129.

The most important in the Army command's view, the reconnaissance, gathered information on the enemy and directed artillery fire. The plane, usually a tractor with a 125 to 200 horsepower engine, carried a pilot and observer and sometimes a radio. It carried enough fuel for a flight of 6 or 7 hours at a maximum speed of 90 miles per hour and a rate of climb of 5,000 feet in 10 minutes.

The combat aircraft was larger, usually a pusher with as much as 500 horsepower, and equipped with two or three light automatic machine guns and a heavier rapid-fire gun with the personnel to operate them. This plane protected reconnaissance aircraft against enemy planes. It could carry heavy loads but lacked speed, maneuverability, and climbing power. Like the reconnaissance aircraft, it could also carry bombs.

A pursuit aircraft was usually a small, single-seat tractor that featured strong climbing power and high speed. The pilot attempted to get above enemy planes and shoot them down with an automatic machine gun.

[43] Holley, 35.

[44] Holley, 37-38, 63.

[45] Holley, 103.

[46] In December 1914, the Chief Signal Officer recommended to the Secretary of War that the appropriation bill for the fiscal year ending June 30, 1916 include a provision for $500. The money would pay tuition for special technical instruction for officers of the Aviation Section. Eventually, the special instruction would allow the officers to take an air engineering course at MIT, the first of its kind given in the U.S. Two officers were sent after the bill passed for the 1915-1916 term. Hennessy, 123-124.

[47] Holley, 103-105.

[48] By April 6, 1917, five flying schools existed, including Mineola, Long Island, NY; Chicago; Memphis; and Essington, PA. Some Army officers learned to fly at their own expense, including Major William Mitchell, assistant chief of the Aviation Section. Hennessy, 177. Goldberg, 11.

[49] In addition to the 1st Aero Squadron, the 3rd, 4th, and 5th became stationed in the U.S. The 2nd Aero Squadron, which already had a company on hand, came to strength in the Philippines. The 6th squadron became stationed in Hawaii, and the 7th in Panama. Goldberg, 11.

[50] Hennessy, 196. Holley, 37-38. Goldberg, 13. Foulois, 142.

Sources do not agree on the exact number of aircraft available to the Army between purchase of the first plane in 1909 till the U.S. declared war on 6 April 1917. Hennessy concluded that the figure was slightly more than 300. Whatever the number of aircraft produced before the war, however, they were inferior and not suitable for combat. During the war, production was far greater. The Signal Corps acquired 13,894 aircraft. When the U.S. entered World War I, the Navy had 48 officers, 230 enlisted (though NACA statistics showed fewer of each), 54 planes, 2 kite balloons, and a dirigible. Hennessy, 196. Bilstein, 33-34.

[51] Hennessy, 197.

A regulation that required officers to wear spurs while flying illustrated the General Staff's attitude toward military aviation. It took a year of war before that regulation was revoked! Frey, 80.

CHAPTER II

THE GREAT WAR AND AFTER, APRIL 1917 - MARCH 1919

The U.S. participated in the Great War for only nineteen months when it ended November 11, 1918. Because President Woodrow Wilson had insisted on strict neutrality before the U.S. declared war, the nation had made few preparations. Thus, the existing aviation organization was completely inadequate to meet the heavy demands placed on it. In short, the War Department as a whole lacked both knowledge and experience to direct the huge air program. Army aviation found itself unable to fulfill the promises that civilian and military leaders made in 1917 about aircraft production. As the war to end all wars, it brought mass destruction and loss of lives. Despite the horrors brought forth, significant events emerged for the good of the country. Perhaps one of the greatest outcomes was the evolution of an organizational structure for military aviation in the U.S. By 1919, military aviation progressed from a section in the Signal Corps to an Air Service responsible to the General Staff and the Secretary of War. Perhaps even more importantly, the war left a legacy of military aviation and experimental engineering. Within a generation, that legacy proved critical to the nation's survival.

Although WWI ended only ten years after the Wrights sold the Army its first aircraft, air power had matured far beyond the General Staff's concept of it as the "eyes" of the Army.[1] Aviation was one aspect of "total war" that developed after 1914. During the fighting, nation-states mobilized their entire resources, and even civilians behind the front lines were vulnerable to attacks by aircraft. Nations could no longer rely on distance and natural barriers to protect them. The Great War showed that in the future air power would be an offensive weapon — a tremendous strike force. Yet, quick demobilization in the U.S. followed soon after the armistice as the nation tried to find peace in "normalcy."

Mobilizing Aviation Production

On the eve of U.S. entry into WWI, the aircraft and aircraft engine industry consisted of eleven companies. Their total capitalization amounted to less than $15 million. Two firms, Curtiss and Wright-Martin, controlled two-thirds of that investment. The industry had some 8,000 workers, skilled and unskilled, but had not built any combat aircraft.[2]

Aircraft Production Board. Within days after the U.S. declared war, NACA suggested that the council of National Defense switch its focus from industrial and economic mobilization to establish a board to deal with the needs of military aviation.[3]

On May 16, the council organized an Aircraft Production Board (given legal status by Congress on October 1, 1917 and renamed the Aircraft Board) to offset the Signal Corp's lack of experience in managing a large production program.[4] The board was to expedite airplane production and suggest sites for airfields, recommend methods of mass producing aircraft, engines, and accessories, and select the manufacturers best qualified to produce them. For example, under the board's supervision, the aircraft industry began mass production of the Liberty engine.[5]

The Aircraft Production Board, as a subcommittee of the Council of National Defense, assumed an advisory role, though the stature of its members enabled the board to exert great influence. The board functioned as a coordinating agency for the Army and Navy. As soon as the U.S. declared war, the Signal Corps began increasing its overall organization, including its Aviation Section,[6] to deal with the increased demands. The Corps added a General Production Division in April with Waldon (a member of the Aircraft Production Board) as chief. All four divisions of the Signal Corps – Administrative, Aeronautical, General Production, and Engineering (the nucleus of the Equipment Division that formed later) participated in aircraft production.[7]

Equipment Division. The Aviation Section already had an engineering department, though a very small one. This organization, located in the Anson Mills Building in Washington, D.C., consisted of several officers and civilian engineers and draftsmen. Henry Souther, an an automobile executive and a consulting engineer in the Aeronautical Division of the Signal Corps since April 1916, headed the division.[8] Captain Virginius Evans Clark,[9] a graduate aeronautical engineer from MIT who served with the Signal Corps Aviation School's Experimental and Repair Department at San Diego, was in charge of airplane design. Charles B. King led the division in engine design.[10]

Occupied with numerous problems, the General Staff within the War Department lacked experience in air matters. The Signal Corps, under pressure to produce results, expanded the Aviation Section and reorganized it into a number of divisions, all of which were individually responsible to the Chief Signal Officer. Additional divisions established took care of the immediate needs for procurement, construction, and training. Most of these divisions had something to do with aircraft, but few were limited to aviation. These activities, of course, included aeronautics since that was one of the integral responsibilities of the Signal Corps. The Chief Signal Officer quickly found himself in the awkward position of supervising two major programs, and aviation was fast becoming much larger than signals. When the Aircraft Production Board and other organizations got involved with aircraft production, the Signal Corp's position became even more awkward.

On May 24, 1917, the Signal Corps created an Aircraft Engineering Division and placed Major Souther in charge. The division included a Power Plant Department and an Aircraft Department. The Inspection Department of the Aeronautical Division was also transferred to this new division. In order to improve coordination with the Aircraft Production board and NACA, Major Souther moved the organization to the Munsey Building where those organizations were already located.[11] He also began organizing and expanding the division to meet the growing needs of the Aviation Section.[12] He later died on August 15.

By June, the Signal Corps had seven divisions, including Airplane, Aircraft Engineering (with four departments for Power Plant, Aircraft, Inspection, Transportation, and Design and Experimentation), Finance and Supply, Personnel, Schools, Construction, and Balloon. The Design and Experimental Department concerned itself with work at Langley Field on testing of airplanes, engines, related equipment, ordnance, photography, and instruments and accessories. The Power Plant Department conducted work related to development of new aircraft engines by manufacturers; the standardization of engine parts such as propeller hubs, magnetos, couplings, and fittings; and special investigations of engine manufacturers outside the scope of the Inspection Department.[13]

On June 18, 1917, Jesse Gurney Vincent[14] (Vice-President of Packard and designer of the Liberty engine with Elbert John Hall, formerly of Hall-Scott Company), opened an Engine Design Section in offices at the Bureau of Standards furnished for the purpose by the Aircraft Production Board. Vincent brought with him from Detroit a complete engineering organization that he had drawn from various automobile factories, including Cadillac, Dodge, Packard, Pierce, and others. Vincent's section immediately began preparing standardized drawings, bills of material, and specifications for production of the Liberty 12-cylinder engine.[15] A plane design section formed later by Major Clark and took temporary quarters in the Airplane Exhibition Building behind the old Smithsonian Institution.

On July 24, 1917, President Wilson signed the Aviation Act, making $640 million in additional funds available for the aircraft program. The appropriation, the largest yet by Congress for a specific purpose, greatly expanded the aviation program. Even the growing aviation organization of the Signal Corps was too small and inefficient to handle the enlarged responsibilities, though it did become the nucleus. To correct this situation, the first major step was the creation on August 2 of an Equipment Division. Management personnel from the Aircraft Production Board came into the Signal Corps to take charge of production. The Equipment Division was actually a greatly expanded Engineering Division that consisted largely of engineers and designers whose duties included procurement, inspection, construction, design, and laboratory work.[16]

The mission of the Equipment Division was to oversee the production and procurement of aircraft, engines, and accessories for the war. Deeds and Waldon, both of the Aircraft Production Board, were designated Chief and Assistant Chief, respectively. Deeds and Montgomery, who succeeded Deeds in January 1918, also became colonels in the Aviation Section. Because the war had made the production of aircraft and engines critical, the Aircraft Engineering Division transferred to the Equipment Division on August 11, 1917.[17] Colonel Deeds proceeded to reorganize the Equipment Division. He added the Finance and Supply Divisions, making a group of four major departments: Executive, Finance, Supply, and Production (under Colonel Waldon.) Engine and plane design now fell under the Production Department's eleven sections: Engine Design, Plane Design, Engine Production, Plane Production, Ordnance and Instrument Production, Balloon Production, Specifications, Electrical Engineering, Wood, Plant Protection, and Planning.[18]

Experimental Laboratories Established

On June 1, 1916, the Technical Aero Advisory and Inspection Board recommended that the Aviation Section establish a permanent experimental and inspection station for its use. The board, created in 1916 to advise the Chief Signal Officer, consisted of a small group of officers and civilians that included Lieutenant Colonel Clark and Henry Souther as members. The board recommended that the field or base be convenient to aircraft factories in New York, New Jersey, Massachusetts, and Ohio. The station was to be situated along the coastal waters but protected against direct attack from the sea and located where the climate would allow a maximum of flying time. In the Army Act of August 1916, Congress directed the Secretary of War to investigate the suitability of the various military reservations for aviation purposes. A special fund was available to purchase additional sites if the military reservations were found unsuitable.

Langley and McCook Fields. When it was formed in 1915, NACA determined to construct a research facility shared jointly with the Army and Navy. Thus, NACA agreed to accept a portion of the site selected by the Aviation Section in December 1916 from fifteen locations studied for its experimental station and proving ground. NACA named the site Langley Field in July 1917 in honor of Professor Samuel Pierpont Langley, former Secretary of the Smithsonian and a pioneer of American aviation. Following approval by Secretary of War Newton D. Baker's approval in December 1916, NACA purchased 1,650 acres for $290,000 on Chesapeake Bay, near Hampton, Virginia. Construction started in the summer of 1917. Among the first constructions undertaken were NACA laboratories and equipment, but NACA did not get its Langley Memorial Aeronautical Laboratory facilities into use until 1920.[19]

The Signal Corps pushed hard for construction at Langley, beginning in March 1917 because the Aviation Section's small engineering organization in Washington, D.C., planned to move there. Soon after the declaration of war, additional money was allotted to cover further work necessary to meet the immediate needs of the field pending completion of permanent improvements. However, the experimental facilities were still in the planning stage in April. Construction began in May. The Signal Corps Aircraft Engineering Division controlled all experimental, testing, and proving work at the aeronautical experimental station. Progress moved slowly because the site was isolated from skilled labor and because of difficulty getting materials, causing the Navy to abandon Langley in favor of Anacostia on the other side of Chesapeake Bay. The Aviation Section of the Signal Corps also decided to relocate away from Langley.[20]

Vincent expressed his frustration and the urgency of establishing an engineering facility in the following paragraph written to General Squier on July 8, 1917:[21]

> *a mammoth engineering department located here in Washington should have been started a month ago and been a reality by this time. A real executive engineering head shoud have been appointed, and provided with the necessary funds to prosecute aircraft engineering and standardization to the limit. By this I mean that he should have been provided with the necessary quarters to house a large engineering organization plus enough floor space to bring together for examination samples of all the more important aircraft that have been doing good work during the last few months.*

Vincent offered a proposal to General Squier:[22]

> *My plan is simply this: organize an Aircraft Engineering Department here in Washington and house it in a building large enough to not only take care of drafting rooms, offices, etc., but also provide enough floor space to make it possible to bring together the more important aircraft and aircraft engines so that they may be carefully analyzed and the best practices copied.*

Vincent further asserted that *"an executive engineer should be placed in charge of this Engineering Department with instructions to go the limit in securing talent for various branches."*[23]

Vincent doubted that the ambitious aircraft production program could work without a proper facility. He raised the issue with Deeds, explaining that he[24]

> decided that this job could not be done in Washington on account of the impossiblity of securing proper quarters, material and skilled workmen. I suggested that the Airplane Engineering Department be moved to Indianapolis Motor Speedway, together with the attached machine shops and putting in any extra equipment that might be necessary to actually build complete airplanes and fly them.

Deeds expressed cautious interest in Vincent's idea, stating his reluctance in having Vincent and Colonel Clark leave Washington.

Vincent continued to press Deeds to relocate engineering away from Washington, but without success:

> "For some reason I did not seem to make myself clear, or at least, I did not succeed in scaring you and Colonel Waldon as badly as I was scared myself."[25]

Upon the return of Captain Howard C. Marmon from Europe where he had been investigating allied engines, Vincent immediately put him in active charge of engine design as his direct assistant. Vincent then focused his attention on improving organizational alignment:[26]

> During the time that Captain Marmon was cleaning up this job at Detroit, I was continuing to give considerable thought to organization matters, with the view of making some helpful suggestions as to how the Engineering Department should be organized to enable us to properly carry out our program.

Vincent enlisted support from Colonel Clark who had returned from Europe with Marmon and had taken up his position in charge of the Plane Design Section. Vincent prodded him:[27]

> I talked these organization matters over with him many times, and he agreed with me that it would be very desirable to bring the Engine Design Section and the Plane Design Section together into one compact organization in order to fix responsibilty and thereby accomplish definite results.

Vincent's efforts to secure the Indianapolis speedway for his proposed department failed to win support from Deeds. But, recalled Vincent:[28]

A little late on, Captain Marmon and I happened to be in Dayton with Colonel Deeds and were given an opportunity to look over the Dayton-Wright Experimental Field. Captain Marmon and I were much impressed with the possibilities of this field as an experimental station and suggested to Colonel Deeds that the Government take it over for that purpose. Immediately upon reture to Washington, Colonel Deeds took the matter up with Lieutenant Colonel Clark, with the result Clark and I were sent out to close arrangements with the Dayton-Wright Co.

With the aviation program already encountering delays, a board of engineers, composed of Colonel Clark, Majors Vincent and Hall, and Captain Marmon, recommended to the Aircraft Production Board that they establish a centralized engineering and experimental facility in Dayton, Ohio. The city, near to Indianapolis, Detroit, Buffalo, Cleveland, Chicago, Pittsburgh, and Washington, was in an area with plenty of skilled labor. Buildings and grounds were available for immediate use. Some critics later charged that Colonel Deeds favored the site because of his aircraft and other interests there.[29]

In September 1917, the Equipment Division presented a formal request to the Aircraft Production Board for a temporary facility at South Field (Moraine) in Dayton. On September 25, the Aircraft Production Board passed a resolution calling for establishing an experimental factory and testing ground, citing production delays caused by the lack of a centralized engineering facility. Engineering was scattered around Washington at the Bureau of Standards, the Smithsonian Building, the Old Southern Railway Building; and in several other cities, including Mineola, New Haven, Dayton, Detroit, Chicago, and Buffalo. The board also noted that construction at Langley Field was slow and that a favorable location was immediately available adjacent to the Dayton-Wright Company. In addition, Wilbur Wright Field was close by for testing and city utilities and skilled labor were readily available. The board recommended that the Chief Signal Officer provide additional temporary facilities. That day, Deeds informed the Dayton-Wright Airplane Company that, pending completion of Langley Field, the Equipment Division intended to extend the laboratory facilities of the Dayton-Wright experimental field by erecting buildings and equipping them at government expense. A committee appointed by the Equipment Division then met in Dayton. Members of the committee included Charles F. Kettering of the Dayton-Wright Airplane Company; Albert Kahn, architect; Colonel Clark, Major Vincent, and Major Hall.[30]

Vincent recalled the meeting in these terms:[31]

Upon our arrival in Dayton, we met Mr. H.E. Talbott, Sr. and Mr. Kettering and went into conference with them. They stated that they would like to accommodate us in this matter but they did not see how they could give up their experimental station because they had contracted with the Government to do the engineering work on the DH-4 and the DH-9 airplanes preliminary to manufacturing those planes. They suggested, however, that they had another field available which would suit our purposes better on account of the fact that it was nearer town and transportation [and] was, therefore, better. Lieutenant Colonel Clark and I immediately looked this field over, which was then known as North Field, and during this same day we arranged with architects to make plans for the necessary buildings.

The committee found North Field attractive due to its building potential and and avialable utilities available. On September 28, 1917, the committee wired their recommendation to Colonel Deeds. The board concurred in switching to North Field, and Deeds wired back his approval to Colonel Clark that same day. The next day, the Chief Signal Officer, Major General Squier, approved the appropriation for North Field.[32]

North Field consisted of approximately 200 acres of level ground located near the center of Dayton. The lease for the field, drawn up on October 1, 1917 with the Dayton Metal Products Company, provided an option to renew. That day, the Aircraft Production Board also adopted a resolution naming the site "McCook Field" in honor of a large family that had fought with distinction in the Civil War. The board awarded contracts for construction of the first buildings at McCook to the Dayton Lumber and Manufacturing Company on October 2. Construction started on October 10, 1917 and moved ahead rapidly with large crews working day and night, seven days a week.[33]

Airplane Engineering Department. To centralize engineering work and fix responsibility, the Equipment Division moved the Engine and Plane Design Sections from the Production Department into an Airplane Engineering Department in October. Colonel Clark headed the new department and became the first commanding officer of McCook Field. Major Vincent became Executive Officer to Clark and directed engineering activities at McCook Field. Other departments for Legal, Foreign Purchases, and Inspection were created within a couple weeks.[34]

By Executive Order of October 13, 1917, all persons employed at McCook Field were exempted from competitive civil service examinations. Thus, from its inception, Government managed McCook Field as a business institution rather than a military post or station, making military routine of secondary importance. McCook Field was established officially as a Signal Corps experimental laboratory on October 18, 1917.[35]

Vincent recalled that[36]

Upon our return to Washington, we recommended that this [McCook] field be rented and the necessary buildings put up to house the Airplane Engineering Department. This recommendation was acted on favorably and I was informed that I should immediately prepare to go to Dayton and take charge of equipping and organizing this field to make it the headquarters for the Airplane Engineering Department.

Colonel Deeds informed me that he wanted me to be the Executive Officer in this department; first, on account of the fact that I have had wide experience in organization matters, and second, because the engine job was ahead of the airplane job and he wanted Colonel Clark to have a free hand to prosecute airplane experimental work.

I did not like this arrangement very much at the time because Lieutenant Colonel Clark, owing to his rank, automatically became commanding officer, and I anticipated nothing but trouble in an endeavor to put this tremendous job through. Lieutenant Colonel Clark, however, assured me that he would co-operate with me in every way and I decided to do my best to put the job through.

As Executive Officer, Vincent left immediately to arrange for temporary quarters in Dayton for the Airplane Engineering Department and to oversee the building, equipping, and organizing of McCook Field's facilities. Vincent arrived in Dayton on October 14, 1917. He regarded his orders according to the following, dated October 17:[37]

the first of many disappointments that I was to encounter. In other words, I was buried under the Commanding Officer, and although I knew that Lieutenant Colonel Clark would have no objection to my going freely to Colonel Deeds and others with suggestions, I had by this time had enough experience to know that delay was bound to result in spite of anything I could do.

Vincent arranged for temporary quarters on two floors of the recently constructed Lindsey Building. Lieutenant H.E. Blood, in charge of organizational matters of the Engine Design Section at the Bureau of Standards, joined Major Vincent within a few days. Captain Marmon arrived in Dayton a few days after with personnel and equipment for that section. Clark remained in Washington with the Plane Design Section until November 5 in order to complete the first drawings of the Bristol fighter, one of the four foreign aircraft selected for construction in the U.S. for release to the Curtiss Company. Headquarters moved to McCook Field on November 5.[38]

After taking command of McCook, Clark cut short Vincent's efforts to have the Dayton-Wright Company standardize the DH-4 and DH-9 aircraft. He also revised some of Vincent's building plans for McCook. Thereafter, Clark and Vincent had many discussions regarding the organization chart but could not reach agreement, especially because Vincent objected to Clark's dual position. During November and most of December, Vincent lobbied to have Deeds move him away from McCook Field and make him production engineer. According to Vincent,[39]

The principal stumbling block was the fact that Colonel Clark appeared in one place as Commanding Officer, and in another place as Chief Plane Designer. This is, of course, against all logical organization arrangements. In this case it was particularly bad because we had two distinct organizations to weld into one to obtain a harmonious whole. I refer to the men in the Engine Design Section and the Plane Design Section which had been brough on from Washington. The men of the Engine Design Section naturally wanted to report to and work under Captain Marmon or myself, while the men of the Plane Design Section naturally wanted to work under and report to Colonel Clark.

On November 1, 1917, the Signal Corps placed all matters pertaining to technical experimental work at Langley and McCook Fields under the supervision and control of the Equipment Division. The division then organized a separate section to handle the work. Initially, the Equipment Division assigned the technical and experimental work at Langley and McCook Fields to the Airplane Engineering Department. But a week later the Equipment Division divided the work between the two experimental laboratories. Langley Field was given responsibility for instruction and experimentation in bombing, photography, radio, telegraphy, and demonstrations of foreign aircraft. McCook Field was given responsibility for engine and aircraft development, installation of cameras on experimental aircraft, and work on synchronizing machine guns.[40]

The first group of buildings at McCook included the engineering and shops building, the final assembly building, the heating plant, the main hangar, the garage, barracks, mess hall, cafeteria, engine test stands, and transformer house. Two original North Field hangars temporarily served as a dynamometer laboratory and an engine assembly building. By December, the Airplane Engineering Department organization began moving to McCook Field. On December 4, 1917, the department's technical personnel reported to McCook Field, and operations officially began, although some personnel remained in downtown Dayton throughout the war. The organization at McCook grew very rapidly as building and installation of equipment continued throughout the war.[41]

One month after the field opened, the Equipment Division reorganized the Airplane Engineering Department and further defined its functions. It also charged McCook with responsibility for designing aircraft, engines, and accessories; for building experimental models; for testing; and for specifications. This restructuring placed responsibility for new and advanced models of aircraft, engines, and accessories in the Airplane Engineering Department. However, once equipment was developed, control turned over to the Production Engineering Department. The Airplane Engineering Department under Major Vincent's direction as chief engineer maintained a separate staff with offices in the Lindsey Building, and he was responsible to the chief of the Equipment Division. Colonel Clark retained command of McCook Field and also reported to the chief of the Equipment Division.[42]

The operation of the Aircraft Engineering Department as established did not please Vincent. During the month, he pressed the Equipment Division to revise the organization. He petitioned Deeds and Montgomery to set up an engineering organization in Washington that would provide centralized control and overall coordination to the aircraft program. The two officers shunted Vincent to William C. Potter, who became the executive between the Airplane Engineering Department and the Equipment Division. After hearing Vincent's arguments, Potter coordinated with Deeds and Montgomery. They decided to give Potter direct supervision over the Airplane Engineering Department and to keep Vincent at McCook Field. However, as chief engineer of the Airplane Engineering Department, Vincent had absolute authority over all experimental engineering work, whether carried on at McCook or elsewhere. In addition, they told Vincent to organize the department at McCook Field along regular civilian lines and put men in charge of various departments with full control over those departments regardless of military rank.

The Equipment Division proceeded to restructure the duties and functions of the Airplane Engineering Department. Major Frederick T. Dickman relieved Colonel Clark of command of McCook Field's military affairs on January 25, 1918. Colonel Clark, in turn, was placed in charge of original design of aircraft. On February 5, however, Major Dickman was relieved of duty, and the military was then directed to report to the chief engineer. Vincent took control of the Aircraft Engineering Department on February 6 as lieutenant colonel and commanding officer of McCook. Vincent, as chief engineer of the department, was placed in charge of

37

organizing all experimental engineering activities at McCook Field. Activities of the Airplane Engineering Department at both McCook Field and the Lindsey Building were combined at McCook. In addition, Vincent commanded McCook Field until the military unit separated in May 1918. The designer of the Liberty engine now had until the end of the war to shape McCook's future.[43]

Meanwhile, the Equipment Division created a Production Engineering Department, with headquarters in Dayton, on January 11, 1918. In late December, Deeds decided to set it up along the lines Vincent suggested but placed Major B.D. Gray in charge.[44] Major Gray, formerly president of the Hess-Bright Company of Philadelphia, established headquarters in the Lindsey Building. The duties of the department included providing engineering information to manufacturers of aircraft, engines, and accessories. The department tested aircraft in production or ready for production, that is, aircraft beyond the experimental and development stages.[45]

On March 18, 1918, the Equipment Division clarified the functions of the Airplane Engineering Department, directing Lieutenant Colonel Vincent as chief engineer to revise and enlarge the organization as necessary to handle the large volume of work and to safeguard the production program. Responsibily for new and advanced models of aircraft and engines designed to meet military requirements remained with the department. After development, responsibility turned over to the Production Engineering Department. The Production Engineering Department only made changes in design necessary for production. The Airplane Engineering Department took control of any changes having a material effect on design.[46]

On April 15, the Production Engineering Department moved from Dayton to Washington, D.C. and established quarters in the Old Masonic Temple. In June, Major Gray resigned, protesting that interference with his authority and the functions of his department made it impossible for him to act effectively. On July 1, 1918, the Second Assistant Director of Aircraft Production, M.W. Kellogg, appointed J.F. McClelland to head the Production Engineering Department.[47]

Liberty Engine

The U.S. successfully developed aviation engines and accessories to aid the allies during WWI. Perhaps one of the greatest contributions was the 12-cylinder Liberty engine designed by Vincent and Hall. The automotive industry, knowledgeable of mass production techniques, produced the Liberty in the first U.S. attempt to standardize an aircraft engine for mass production. Successful development of the Liberty boosted the nation from a position of inferiority in aircraft engine technology to superiority in less than a year.[48]

Vincent's and Hall's hard work and dedication enabled them to produce the design of the Liberty. Although the engine was reputedly the result of their six-day design session in the Willard Hotel in late May and early June 1917, it actually represented the culmination of years of experience combined with skill and insight. The Liberty design, a departure from the heavy, durable aircraft engines of that time, offered light weight and great power.[49]

Originally designated the U.S.A. Standardized Aircraft Engine, the engine's popular name was suggested by Admiral Taylor and it immediately caught on. The Liberty came in a series of 4-, 6-, 8-, and 12-cylinder models, each with a 5-inch bore and a 7-inch stroke, developed for interchangeability of parts and ease of production. The engine remained in active use by the Army Air Corps throughout 1936, and had other applications even after WWII.[50]

The Production Dilemma. By the time the U.S. entered the war, air weapons had already reached a sophisticated level in Europe. German dirigibles bombed Paris, London, and other cities. Highly specialized aircraft types such as including bombers and fighters had evolved along with related tactics and strategy. Air services in Europe included multi-engine bombers armed with machine guns to use against pursuit planes. Opposing fighters dueled for control of the air. Bombing, strafing of ground troops, and aerial reconnaissance and photography were widely accepted tactics.[51]

In 1917, the aviation industry was only beginning to evolve. America was not only behind in state-of-the-art engine development, but also in aviation instrumentation. Dials, gauges, and other accessories for airplanes were unfamiliar to American manufacturers. Other problems also had to be solved. Spruce was needed for wing spars and yards of fabric to cover airframes. Because ordinary motor oil congealed at high altitudes, castor oil became a necessity. The U.S., however, did not have enough castor beans to produce the millions of gallons of oil required.[52]

After declaring war, the nation faced the complicated task of aircraft production. Mass production in the U.S. traced back to Eli Whitney. Henry Ford and others had applied standardization to produce automobiles and other commodities, but the techniques were not common or necessary in aviation production. Once war erupted, the aviation industry had to build machine tools and adapt mass production techniques quickly. Production of aircraft and aircraft engines had to be accomplished while competing for manpower, equipment, supplies, and resources. Skilled manpower and materials were in short supply, and expert capability in the Army was limited. The emphasis on engine power and reliability with extreme lightness of construction created severe demands. These problems were exacerbated by the distance separating the experienced technicians in Europe from the U.S.[53]

Prior to April 1917, only four U.S. companies had successfully produced aircraft engines. These firms included Hall-Scott Motor Car Company of Berkeley, California; Curtiss Aeroplane and Motor Corporation of Buffalo, New York; Wright-Martin Aircraft Corporation of New Brunswick, New Jersey; and General Vehicle Company of Long Island, New York. The capacity of these companies was quite limited. Their engines, built for durability rather than lightness, produced only enough power for training aircraft. The war brought two propulsion requirements: an enormous increase in the production of engines for training aircraft and development of a new engine for combat aircraft.[54]

After the U.S. entered the war, the allies hastened to provide their technical knowledge. Within three weeks, large missions arrived in Washington from both France and England. They felt that the U.S. could help most effectively by sending a powerful air force to support the hard-pressed Allies on the western front in 1918. In a cable to President Wilson, received on May 26, 1917, Premier Alexandre Ribot of France made a proposal that became the basis of the U.S. wartime aviation program. Ribot asked for an American flying corps of 4,500 planes, 5,000 pilots, and 50,000 mechanics in 1918 with 2,000 planes and 4,000 engines constructed monthly beginning in January 1918. During the first six months of 1918, the allies needed 16,500 aircraft and 30,000 engines, not including training planes and engines. The magnitude of the program was staggering. In three years of war, France had not produced the number of planes it wanted the U.S. to produce in one year. Acting under great pressure, a group of officers headed by Foulois drafted a program within in a few days based on Ribot's cable. Their proposal called for producing 22,625 aircraft and 45,250 engines.[55]

To meet Ribot's request required the U.S. to produce far more planes and engines than previously proposed. For example, NACA recommended in April 1917 that the U.S. produce 3,700 aircraft in 1918, 6,000 in 1919, and 10,000 in 1920. The War Department set aside NACA's proposal in favor of the program based on Ribot's cable. The Secretary of War Newton D. Baker approved the program before formal action by the General Staff and submitted legislation to Congress. Reports of the air war in France had fanned interest in having a dominant American air arm, and the press and public were enthusiastic. Inspired by the public's enthusiasm, Congress rushed through an appropriation of $640 million for aeronautics in 15 days. President Wilson signed the act on July 24, 1917. The appropriation included $125 million for aircraft and $240 million for engines.[56]

The Propulsion Challenge. Major Bolling led the largest American commission to go to Europe during the war. A graduate of Harvard Law School and general counsel for United States Steel Corporation, he went abroad to negotiate aviation agreements on patents, aircraft manufacturing, and engine production. The Bolling commission arrived in England on June 26 and visited France and Italy before disbanding in mid-August 1917.

Each member filed a separate report. By the time of Bolling's final report, the development of an American-designed series of standard aircraft engines was already in progress. In fact, the 8-cylinder engine had been tested and the 12-cylinder model was in final testing by the time the Bolling commission disbanded.[57]

In this period, four basic types of American aircraft engines were active: radial, rotary, vertical line, and V-type inline. The radial engine's cylinders were arranged like a fan, but the cylinders remained stationary while the crankshaft revolved. Since the radial engine was difficult to cool with water, most were air cooled. Although the bulk of the engine caused great head resistance, the radial configuration produced a compact, lightweight engine.[58]

The rotary engine was similar to the radial in that the cylinders encircled the crankshaft. The crankshaft on the rotary was fixed and the cylinders turned around it. This engine was light and easily cooled by air, but the resistance of the air to the rotation of the cylinders absorbed much of the engine's power. Because lubricating the rotary required use of an oil-gasoline mixture, the engine used excessive amounts of oil. To increase power, radial and rotary engines added cylinders. Unfortunately, that also increased complexity and maintenance difficulty.[59]

The vertical inline engine's cylinders stood in a row directly above the crankshaft as in an ordinary automobile engine. This engine was prominent in the development of aircraft engines. A 1912 list of 112 aeronautical engines included 42 vertical types. The four-cylinder vertical vibrated and lacked power, making it an engine aviators disliked. In Germany, Mercedes and Benz brought the six-cylinder model to a high degree of perfection, but the vertical remained unpopular in the U.S.[60]

The V-type inline engine's cylinders extended upward at an angle from the crankshaft in two banks. Radial and rotary engines were lighter than inlines. However, due to their limited cooling capacity and speed of rotation, they could not develop sufficient power for military applications. The most satisfactory military engine from a power-to-weight ratio, therefore, was the V-type inline, the basic design of the Liberty.[61]

During WWI, the U.S. had to mass produce aircraft engines with greater horsepower than ever before. Two choices were available: reproducing the most effective allied engines or designing and producing an American engine. Production of European engines in American factories would have been difficult because foreign engines were handmade, parts were not interchangeable, and they used the metric system. By 1917, the Wright-Martin Company had spent almost two years and $3 million producing the Hispano-Suiza engine, and achieved only limited production. Similar results were experienced producing the Le Rhone, the Gnome, and the Bugatti engines. The allies developed and manufactured sixty different engines, a costly approach that resulted in low production and serious shortage of parts in the field. The Germans, concentrating on only five types of engines, outproduced them.[62]

Requirements called for designing and building an American engine with great horsepower that could be produced in quantity. By producing standardized engines with interchangeable parts, the U.S. could minimize the difficulties of supplying and repairing engines at a battlefront 3,000 miles away. The major requirements included maximum power and efficiency with minimum weight; capacity to run at maximum power and speed for a large percentage of its operating time; economical fuel and oil consumption; and in order to produce a reliable engine in the shortest amount of time, off-the-shelf components. To overcome the expense and technical problems involved in developing engines of 400 to 500 horsepower required the U.S. to pool its resources. Furthermore, Government control of U.S. production would enable engineers to exploit foreign and domestic experience.[63]

Design and Qualification. Deeds acted as the catalyst in the decision for the Liberty engine in 1917, but Vincent and Hall also participated.[64] Since 1914, Vincent had been experimenting with several types of 12-cylinder aircraft engines with 225 horsepower. Vincent's design of 905 cubic inches, tested in December 1916, proved too heavy for aircraft use. Still, he had amassed a great deal of information and experience, and had built up an experienced experimental engineering section at Packard. Meanwhile, Hall completed an experimental 12-cylinder engine, the A-8, designed to produce 450 horsepower. Hall's engine was ready for final testing when the Navy called him to Washington in May 1917. By then, the French and British missions were in Detroit surveying the industries there. Vincent learned a great deal about engine developments in Europe from them.[65]

After discussions with the British and French missions, Vincent realized that quick, decisive action was necessary to avoid the same mistakes made by the allies in engine production. He received permission from Packard Company's president to go to Washington and present his ideas about a standardized line of aircraft engines to the Aircraft Production Board. In May, Vincent discussed his ideas with Coffin, Deeds, and Waldon, claiming that the excess weight of his Packard aircraft engine could be reduced without sacrificing reliability. Deeds, who had been thinking along the lines of a standard engine, realized that the government would have to be the driving force behind such a project. He asked Vincent to work with Hall in designing a standardized engine.[66]

Vincent and Hall, who knew each other only by reputation, met with Waldon in Deeds' suite at the Willard Hotel on May 29. Deeds impressed them with the need for speed and, in order to assure rapid production, cautioned them to use only state-of-the-art components in the new engine. They immediately got to work. With support from Washington's Society of Automotive Engineers, the Bureau of Standards, members of the French mission, and a draftsman, Hall and Vincent proceeded to block out the drawings. They reported their progress to a conference of the Aircraft Production Board and the Joint Army-Navy Technical Board on May31, receiving approval to complete the drawings. On June 1, two layout men arrived from Detroit to help. The five men worked through the afternoon, and at midnight, Hall and Vincent met with the two boards again to show their drawings and explain their plans for the engine. The boards approved building five 8-cylinder and five 12-cylinder models, and urged that the first 8-cylinder be produced as quickly as possible. Vincent sent the layout men to Detroit to work on detailed drawings.[67]

On June 6, Packard's president in a meeting with Vincent, Deeds, and Waldon agreed to pioneer the standardized engine and to finance it until the Government could reimburse the company. He also loaned Vincent to the government for three months and gave top priority to the engine project in the Packard plant. On June 7, Vincent and Hall traveled to Detroit to supervise production of the experimental engines at the Packard plant.[68]

Foreign experts had agreed in May 1917 that the 8-cylinder engine, rated at 225 horsepower, would meet the requirements at the front in the spring of 1918. The 12-cylinder model, rated at 330 horsepower, was to be the engine for 1919 and 1920. However, events in the air war moved so rapidly that within 90 days the experts called for a 12-cylinder model with 400 to 450 horsepower in 1918. The design of the Liberty allowed this escalation. Packard draftsmen pushed the engine work rapidly, working weekends. After completion of the 8-cylinder drawings, Vincent took some 25 draftsmen furnished by Dodge, Packard, Cadillac, and Pierce Arrow to Washington to work on the 12-cylinder drawings. They sent the completed drawings to various automobile companies to have tracings made, and some 300 draftsmen worked for a week on the tracings. All drafting work ended by June 15, 1917.[69]

Vincent shipped a wooden model to the Bureau of Standards on June 16. By then, various companies had orders in for 8-cylinder parts. Following assembly and testing in the Packard plant, delivery of the first sample 8-cylinder engine to the Bureau of Standards took place on July 3, 1917. Only a month had passed since its conception. During that month, Deeds had submitted the design to engineers and manufacturing experts at various automotive and aviation companies for review, and all approved the design.[70]

In early July, work completed on consolidating and numbering the drawings and checking and correcting tracings, bills of material, limits, and related activities. Late in July, Hall ran the first standard 8-cylinder engine under its own power in Detroit. Vincent arrived at the Packard plant on July 25 to conduct testing. The engine experienced no vibration up to 2,000 revolutions-per-minute. Mounted on a truck and fitted with a propeller, they tested the engine for an hour and a half with no trouble. At month's end, they demonstrated the engine at the Bureau of Standards in Washington to Deeds and Waldon. Successful testing led Vincent to recommend immediate production of the engine. He received permission. Following further testing, the engine was disassembled on August 6 and found in excellent condition.[71]

Only 68 days after conception, the standard-detail drawings for the L-8 and L-12 were complete; the standard bills of material and the standard material specifications for both models were complete; the final construction drawings for both engines were within two weeks of completion; and the general designs were approved for manufacture. On August 13, the first L-12 began testing. By the end of August, the engine had completed its 50-hour testing in an elapsed record time of 55 hours and judged satisfactory.[72]

In November 1917, Navy Lieutenant Harold H. Emmons,[73] chief of the Engine Production Section in the Equipment Division, directed that a production engine from the Packard plant be tested to destruction. Vincent, almost jealously involved with the Liberty, protested his exclusion from this testing. In turn, Deeds directed Vincent and Hall to observe the testing, a quick, conciliatory action characteristic of his wartime leadership. Tests on two engines, numbers 5 and 12, modified and strengthened to meet the demands for more horsepower, ended in February 1918. This accelerated development of the Liberty engine was characteristic of how the war stimulated technical progress. Within a year of the first design meeting in May 1917, more than 1,000 engines were produced. The Liberty engine proved an unqualified success.[74]

The Liberty engine's first flight occurred with L-8 number 3 on August 29, 1917. The 20-minute flight took place in Buffalo, New York in an LWF aircraft. After a subsequent flight, an Army officer noted it was the first aircraft he had flown that seemed to have a surplus of power. Flight testing of L-12 engines began in October 1917. The first L-12 flight occurred in a Curtiss HS-1 flying boat for the Navy on October 21. The first flight in an American-built DH-4 occurred on October 28, 1917. The first production L-12 engine was delivered to McCook Field on Thanksgiving Day 1917.[75]

Production and Characteristics. The Aircraft Production Board, though advisory, investigated manufacturers and recommended which companies should receive production contracts for the Liberty engine. The Army, as the responsible contracting agent, furnished the Navy and allies with engines from those produced. The contracts used for procuring engines used cost-plus-fixed-profit margins. The cost to build an engine, as set by a group of manufacturers not involved in the program, was $6,087 plus a profit of $913 with a bonus for economy in production. The bonus consisted of a split of 25% to the manufacturers and 75% to the government of any reduction in the $6,087 cost. The cost was $17.50 per horsepower ($7000 per engine divided by 400 horsepower) compared to Rolls-Royce's $23 per horsepower. In December 1917, the cost of building an engine, based on quantity production, was set at $5,000 including a fixed profit of $625, and all contracts were amended accordingly.[76]

Problems in producing the Liberty engine plagued manufacturers The requirement for interchangeable parts led to strict inspections, and caused the government to set very rigid tolerances for safety. The margin of safety had been reduced to a minimum to decrease weight. Adherence to specifications became even more important after enforcing engine modifications to produce greater power. However, inexperienced inspectors slowed production, especially early in the program, because they did not know when to permit deviations from the basic design. Shortages of natural resources also caused problems. Lumber and metals were in short supply. Tools, jigs, gauges, components like spark plugs, and related items were scarce in the early months of production. It became necessary to draft skilled workers and train women entering the work force. A shortage of coal in the winter of 1917-1918 forced some factories to shut down.[77]

A steady stream of design changes caused costly production problems. Manufacturers continually updated their production engines according to modifications resulting from tests to increase the horsepower of the L-12 model. Similarly, changes to blueprints resulted in delays and excess scrap. These changes averaged 100 per week and affected 25 percent of all engine parts.[78]

The basic premise of the Liberty design was interchangeability of parts. All models used as many identical parts as possible including the same cylinders, pistons, and numerous other parts like crankshafts and crankcases. Although the smaller engines were sound (the L-8 vibrated excessively), the wartime need for power pushed the smaller engines into the background. Only the 12-cylinder engine was used extensively. Manufacturers built two 4-cylinder engines, 52 6-cylinder engines, and 15 8-cylinder engines. They also built 20,478 12-cylinder engines; 13,574 through the armistice and 17,935 by December 31, 1918. Government let contracts for a total of 56,000 L-12 and 8,000 L-8 engines, though they canceled most of them after the armistice. Overall, 32,420 engines were produced by the armistice; almost half were Liberty engines. More than a quarter were OX-5s, a Curtiss 90-horsepower engine used in training aircraft, and another 8,000 were foreign types, mostly Hispano-Suiza and Le Rhones.[79]

The weight of the Liberty 12-cylinder varied depending on accessories and equipment. The basic engine, dry and without a radiator, weighed 786 pounds, but fully equipped for flying it weighed over 900 pounds. Other characteristics of the L-12 included:[80]

Type:	V 45 degrees	Weight per hp:	2.11 lbs
Cylinders:	12	Fuel per hp hour:	0.509 lbs
Horsepower:	400	Oil per hp hour:	0.037 lbs
RPM:	1,800	Average cost:	$4,000
Bore & stroke:	5"x7"		

The original 350-horsepower L-12 engine was a fine, economical, dependable engine. By May 1918, the 12-cylinder Liberty produced 450 horsepower at 1,800 revolutions per minute. At a weight of 825 pounds, that gave 1.8 horsepower per pound, making the Liberty the lightest engine per horsepower per pound in the world.[81] Criticism of the L-12 engine, when horsepower had to be increased to the 450 to 500 horsepower range, focused on the scupper oiling system. To achieve the increased horsepower, a forced-feed system had to be designed and the engine's bearings, crankshaft, and other parts had to be strengthened.[82]

Vincent and Hall designed the Liberty, but the engine reflected the ideas and experience of many people. The creators drew not only from their own proven designs but also from the designs of other manufacturers. Vincent noted:[83]

Every feature going into the Liberty motor had been thoroughly proved out in Europe and also by experimental work in this country. I had personally spent two years at the Packard Factory developing the improved type of valve action which was used in the Liberty motor, as well as light steel cylinders, the water jacketed intake headers, the two part box-section crankcase, and so on through the list of features, which are now well-known as being important features of the Liberty motor.

In another account, Vincent listed the features that he contributed as follows:

- crankcase construction split on the center line with the bearings carried between the two halves and through bolts running from top to bottom.
- steel cylinders of the Mercedes type of construction, but designed for rapid production.
- camshaft and valve rocker arm construction.
- intake header and carburetor arrangement including means for heating the intake header.
- 45-degree angle of cylinders.
- water pump design, location and drive, including self-takeup on the stuffing box.
- connecting rods and bearings.
- oiling system as finally adopted including full pressure feed, no grooves in the bearings and tripple oil pump to accomplish dry crankcase.[84]

Many Hall-Scott Company features were included in the Liberty, based on experience producing the A-5, A-5a, A-7, A-7a engines, and the A-8 designed to be a 12-cylinder engine of 450 horsepower. The A-8 and L-12 had many similarities, including:

- overhead cams
- individual and interchangeable cylinders
- the same 5-inch bore and 7-inch stroke in both engines

- propeller hub and bolts
- direct drive, nongeared timing
- seven bearing crankshaft
- crank throw bearing centers
- ignition
- distributor head
- bevel gear on camshaft driveshaft
- piston rod sections
- submerged oil pump in sump

Other Hall-Scott features included:

- heavy duty, die-cast, aluminum-alloy pistons
- enclosed camshaft drive
- method of drawing water from the exhaust valve side of the cylinder
- propeller flange drive
- crankshaft and bearing diameter proportioned to the horsepower
- all the dies used in producing the first Liberty engine

Hall also contributed the direct drive feature.[85]

Features taken from other engines included the following:

- 5-inch by 7-inch cylinder of Curtiss and Lorraine-Dietrich, as well as by Hall-Scott
- cylinder design of Mercedes, Rolls-Royce, and Lorraine-Dietrich
- camshaft based on Mercedes, Hispano-Suiza, Rolls-Royce, Renault, Fiat, and Hall-Scott
- 45-degree angle of Renault and Packard
- Delco ignition used in automobiles
- forked connecting rods used DeDion, Cadillac, and Hispano-Suiza
- cranshaft design of Mercedes, Rolls-Royce, Curtiss, Renault, and Hall-Scott.

Vincent's crankcase was similar to that used by Mercedes and Hispano-Suiza. The original and redesigned lubrication systems were similar to Rolls-Royce and features of the Hispano-Suiza. Hall's propeller-hub design was similar to the Mercedes. The Liberty used a conventional centrifugal water pump and a Zenith carburetor.[86]

In January 1918, work started on gearing the propeller of the Liberty engine to provide more power. The engines produced for the war were direct drive, a method of propulsion cheaper to produce and lighter and easier to maintain than a geared method. The primary Liberty-equipped aircraft for use in France by the U.S. was the

DH-4, and the direct-drive L-12 was adequate for this aircraft. Using it in larger, multi-engined aircraft and flying boats, however, required more power. Operating the engine at higher engine revolutions-per-minute produced more power. Therefore, gearing that lowered propeller revolutions-per-minute was necessary, thereby increasing efficiency.[87]

The Liberty was a tremendous leap forward and the allies encouraged its production, although it was not perfect. Criticism of the Liberty engine occurred, stemming from the attacks on the aircraft program by the famed sculptor of Mount Rushmore, Gutzon Borglum. His attacks also unleashed criticism of the engine, including several features: the 45-degree cylinder arrangement (it worked); the battery ignition system (better than contemporary magnetos); and the engine oil sump cover (it worked.) Other criticisms of the Liberty engine claimed it was not suitable for single-seat fighters (not true); was not fast (fast enough); was hard to cool (improvements were possible); used excessive fuel (average); and used excessive oil (oil quality was the real issue.)[88]

Vincent's believed that the Liberty engine required castor oil for peak performance. Before U.S. entry into WWI, high-powered internal-combustion engines, especially rotary engines, used castor oil because of its greater viscosity, high flash-and-fire tests, and great penetrative qualities. Castor oil was the only satisfactory lubricant for the Liberty until development of a mineral oil, called Liberty Aero Oil. Lacking castor beans, the U.S. formed a Castor Oil Board in October 1917 to promote production. Britain offered to provide up to 3 million gallons overseas, but the U.S. still needed to find another 3 million gallons. The board laid out a detailed production program, offered incentives, and distributed seeds. Problems such as ignorance, late planting, poor seeds, inferior machinery for harvesting and hulling, drought, a shortage of labor, worms and mold undermined production.[89]

After months of experimentation, however, an acceptable mineral oil was developed in November 1917. Liberty Aero Oil, produced at a cost of 75 cents per gallon (only one-fourth the cost of castor oil), proved effective in the Liberty 12's high temperatures and bearing pressures. Pilots who were using more than 20 types of oil, however, were difficult to convert:[90]

> *Much opposition was experienced at first and the department was obliged to resort to scheming. Whenever the flier or motor maker refused to believe that Liberty Aero Oil was equal to or better than his favorite brand, this oil was put into containers from which he supposedly drew his pet supply. The flier would be requested to state the efficiency of his lubricant at the end of the flight and when his unqualified praise was given, he was advised that Liberty Aero Oil had been used instead of his pet brand. It was not long until Liberty Aero Oil was successfully introduced to and universally praised by the government aeronautical service.*

The DH-4 Aircraft. The combat career of the L-12 was tied to the American-built DH-4, an aircraft that remained in the Air Corps inventory through 1931 (not because it was such a wonderful aircraft, according to General Arnold later, but because it was all the Air Service had.) The decision to build an American engine did not extend to building an American combat aircraft. When the U.S. entered WWI, the only available aircraft were built by Curtiss, Glenn Martin, Standard, and Lowe, Willard and Fowler (LWF.) These aircraft were powered by Curtiss 90- and 160-200 horsepower engines, Thomas 140-horsepower, Hall-Scott 130-horsepower, and Sturtevant 140-horsepower engines. After declaring war, the initial U.S. effort aimed at building a fleet of trainers, mostly Curtiss JN-4 biplanes, to provide pilots.[91]

Meanwhile, the Bolling commission gathered information in Europe to recommend a course of action for the U.S. Colonel Clark, a member of the commission, described three types of aircraft operating at the front: observation, combat or pursuit, and bombers. He believed they required engines ranging from 235 to 520 horsepower. The Bolling commission concluded that the French and British combat and pursuit aircraft were adequate and that the U.S. could produce them. Thus, the U.S. purchased fighter aircraft from its allies, mostly French Spads and Nieuports, and built other foreign aircraft under contract. Plans to produce four included the DH-4 combination reconnaissance aircraft and day-bomber, the Handley-Page night bomber, the Caproni bomber, and the Bristol fighter. The U.S. role, then, was to produce the larger observation and bombing aircraft, types that changed more slowly. The Liberty engine suited these aircraft, but the General Staff's belief that observation aircraft were primary in war also affected the decision.[92]

The DH-4 aircraft fit both criteria. The sample DH-4 reached New York on July 18, 1917 and immediately sent to Dayton for redesigning to take American machine guns, instruments, other accessories, and the Liberty engine. Although the first American-built DH-4 was ready to fly on October 28, 1917, much work remained before the aircraft was useful. The American DH-4, a good machine that gave excellent service, was being called a "flaming coffin" even before it first flew in France on May 17, 1918. However, no greater percentage of DH-4s were lost in flames than any other type of aircraft at the front even though self-sealing gas tanks were not installed in the aircraft until October 1918. Production of the DH-4s reached a total of 4,846 aircraft, of which 3,431 were completed and shipped from factories by November 11, 1918 and 2,297 were floated from ports of embarkation. One L-12 engine powered the DH-4, supported by another plus spares. Enemy action destroyed thirty-three DH-4s, 14 percent of the total lost by American squadrons, but the type claimed 59 victories.[93]

Many aircraft used the L-12 engine, but only the DH-4 saw service in France. Considerable effort went into the Handley-Page bomber which was adapted for production. A set of drawings came from England in

August 1917, but during the winter two new sets were sent that altered almost every part. Because of its large size, assembly of Handley-Page parts manufactured in the U.S. took place in England. Manufacturers sent 101 sets of parts for the Handley-Page overseas between July and October 1918. Unfortunately, the armistice intervened and they never reached the front.[94]

Post-War Service. By August 1919, 612 DH-4s were shipped back to the U.S. and distributed around various flying fields, causing a surplus of training aircraft. The war's end also left a huge surplus of L-12 engines. In October 1919, the Army had 11,871 Liberty engines; 2,773 in service and the rest in storage. In August 1919, the Air Service announced a policy under which serviceable Liberty engines would be sold to U.S. citizens for commercial and civil aeronautics and to educational institutions. The Post Office Department and several schools bought some of them.[95]

Development of an air-cooled version of the L-12 began in August 1923. Air cooling improved flying qualities by reducing weight, streamlining, improving pilot vision, reducing noise, and providing smoother operation. Air cooling also reduced costs and eased maintenance. Air-cooled engines were 141 pounds lighter and produced 436 horsepower at full throttle -- power as good as the water-cooled version. Because few were produced in 1924 and 1925, the cost of these engines, geared and inverted, was over $8,500 each, about twice the cost if produced in large lots.[96]

In the end, the long life of the Liberty engine proved a mixed blessing. The engine endured in a world where its attractiveness depended on its low cost resulting from large surpluses. The L-12 was so far ahead of its time that it took several years for manufacturers in the U.S. to surpass it. During its heyday, it was the bulwark of the Air Service, the Air Mail Service, and commercial aviation. Mass availability of unused engines made it unnecessary to update and recondition used engines until 1929. Existence of these engines tended to restrict development of new service aircraft. By 1924, new engines were on the market or in development that were superior to the Liberty in performance causing concern that the Air Service would stagnate at 1918 technology. In 1924, the Air Service projected a 26-year supply due to a surplus of 11,810 engines, clearly a drag on engine development. Selling them would have benefited commercial aviation which was then built around old Curtiss trainers since high-powered engines would have forced production of a more satisfactory commercial airplane. By 1927, when the Air Corps estimated that it still had a 20-year supply of L-12s, the water-cooled Curtiss D-12 with 435 horsepower had become the standard for pursuits. Even commercial companies refused to take the Liberty engines from the government, especially because air-cooled versions offered greater profit. Finally, on July 1, 1929, the Air Corps prohibited the use of Liberty engines in any new aircraft.[97]

Speedboats continued to use the L-12 long after their unsuitability for aircraft. During prohibition, rum runners used boats with Liberty engines purchased from junk dealers to outrun Coast Guard boats, forcing the Coast Guard to equip its boats with L-12s. In 1929, the Army tested the Liberty in armored cars modified from WWI Christy tanks. The Christy took the Liberty to war on the eastern front in 1941, driving Russian tanks against the Germans. The British also used the American Liberty engine, built under U.S. license, in some of their tanks in WWII.[98]

Air Service

By the spring of 1918, the vaunted aviation program of 1917 was in serious trouble. Production had not begun to approach the goals announced, and public optimism was disappearing rapidly. Public optimism dwindled following charges that the agencies handling the program were inefficiently organized and administered. Criticism grew bitter. By April 1918, several official investigations were underway, including a Senate inquiry. On April 10, 1918, the Senate recommended that the aircraft production program be removed from the Signal Corps and placed under the direction of a person who had no business interest in the production of aircraft or equipment. A sweeping reorganization of the War Department's whole aeronautical structure followed.[99]

The men in charge of aircraft production and the organization to promote it, most of whom came from outside the nascent pre-war aircraft industry, became the target of criticism. The smaller airplane manufacturers blamed the problems with aircraft production on the "Detroit Gang" of automobile men. For example, Grover C. Loening, one of the few aircraft engineers in the U.S. at the start of the war, charged that his company had been treated unfairly by the big businessmen. Loening argued that a grave mistake had been made in depending entirely on producing foreign combat aircraft that were not originally designed to use the Liberty engine. Loening believed that American engineers, though youthful, were as capable of designing combat aircraft in 1917 as their counterparts in Europe. These criticisms, however, did not recognize U.S. successes, especially considering that the U.S. was at war for only 19 months. During that period, the U.S. produced more aircraft and engines than did England and France.[100]

Deeds received the severest criticism from the investigators. The Delco Company, of which he had been president, produced equipment for internal combustion engines, including ignition, lighting, and starting systems. When the war started in Europe, he helped organize the Dayton-Wright Company. Before the U.S. became engaged, Howard Coffin, Chairman of the Council of National Defense, called Deeds to Washington to serve on the short-lived Munitions Standards Board. In May 1917, Deeds became a member of the Aircraft Production

Board, and then in August 1917, chief of the Equipment Division. Commissioned a colonel in the Signal Corps, he became one of those criticized as "arm chair colonels."[101]

As a result of the public outcry and the resulting confusion, in May 1918 President Wilson appointed Charles Evans Hughes, a former Justice of the Supreme Court, to investigate the situation. The Hughes report, not released until October 1918, recommended that they court-martial Deeds. Deeds, relieved of membership on the Aircraft Board on May 22, 1918 along with Colonel Montgomery and ordered to report to the Attorney General for investigation, was accused of acting as confidential advisor to his former business associates and of issuing false and misleading statements about aircraft production. Hughes also recommended criminal prosecution of three other officers: Lieutenant Colonel Vincent, Major George W. Mixter, and Lieutenant Samuel B. Vrooman, for transacting business with concerns in which they had financial interests. A special War Department board reviewed the evidence against Deeds, but in January 1919, Secretary of War Baker recommended against prosecuting him because he had disassociated himself from business organizations holding government contracts. Legal action was not taken against anyone after Secretary Baker's statement.[102]

Meanwhile, organizational changes accompanied the investigations, beginning in April 1918. As the first step, the War Department placed John D. Ryan, a prominent banker and president of the Anaconda Copper Company, in charge of aircraft production. Next, President Wilson issued and executive order on May 20. Under the Overman Act of that date, the president was given authority to redistribute functions within the executive department during the war and for six months afterwards. The president removed aeronautics from the Signal Corps, permitting Major General Squier to concentrate on administering signals. The president also approved creation of two independent branches of the War Department, a Bureau of Aircraft Production and a Division of Military Aeronautics (later referred to as the Department of Military Aeronautics.)

The Bureau of Aircraft Production and the Division of Military Aeronautics were independent of the Signal Corps and under the direct supervision of the Secretary of War, but the rules and regulations of the Aviation Section carried over to the new organizations. BAP was assigned exclusive responsibility for production of aircraft, engines, and equipment for the air arm. DMA was to conduct training and operations of the flying and ground forces for duty in Europe. The Bureau of Aircraft Production included the former Equipment Division and part of the Supply Division of the Signal Corps. On May 24, 1918, the War Department officially recognized DMA and BAP as constituting the Air Service; however, no chief was appointed to coordinate their activities. On May 29, Ryan assumed control of the Bureau of Aircraft Production. Brigadier General William L. Kenly became director of the Division of Military Aeronautics. DMA established a Technical Section under Lieutenant Colonel Thurman H. Bane.[103] The exact division of functions between the DMA and BAP in designing and engineering was to be worked out as experience evolved.[104]

In May 1918, Ryan began negotiating to hire Charles W. Nash, president of Nash Motors Company and former president of General Motors Company. Ryan wanted Nash to centralize the engineering work of aircraft production and to coordinate engineering and production. Nash at first only agreed to review the situation. In June, he reported that the various departments were pulling in different directions, often at odds even within the same department. He recommended that all branches of aircraft production be located at Dayton because of its facilities and central location. He also recommended that aircraft engineering and production be unified under one head with full power to carry out whatever program Washington directed. On June 24, shortly after receiving Nash's assessment, Ryan moved to centralize BAP's engineering and research work by creating an Engineering and Research Division. That division was to include the Airplane Engineering Department, the Science and Research Department,[105] the Technical Information Department, and the Production Engineering Department. The new division was to operate under the Second Assistant Director of Aircraft Production, M.W. Kellogg. Nash also agreed to take the position that Ryan had offered, becoming First Assistant Director of Aircraft Production in charge of engineering and production on July 17, 1918. Nash proceeded to coordinate the activities of the Engineering Department, the Production Department, and the Technical Section of the Division of Military Aeronautics with the intention of centralizing these functions at McCook Field.[106]

Meanwhile, Ryan and General Kenly made an agreement on June 6 relating to the duties of their respective organizations. The agreement, however, failed to clarify the exact jurisdiction of their organizations relative to technical and engineering problems. A second agreement was hammered out in late July that coordinated the work of BAP's Airplane Engineering Department and DMA's Technical Section in order to preclude duplication, delays, and misunderstanding in handling new designs of aircraft and engines. The agreement located both BAP's Engineering Department and the DMA's Technical Section at Dayton. The agreement also defined the responsibilities of the two organizations and established procedures for each, including the roles of the Production Engineering Department and the Airplane Engineering Department.[107]

On August 7, 1918, Nash informed Ryan that his plan to centralize as quickly as possible at Dayton had general support. He also reported that, even though the Technical Section of DMA and the Engineering Department of BAP were already working in harmony, he intended to unite the two organizations. The Technical Section soon relocated to Dayton from Washington.[108]

The attempts by DMA and BAP to establish a workable arrangement relative to technical and engineering responsibilities proved inadequate. Actually, the original plan had envisaged the eventual consolidation of the two new agencies under a single director. Thus, on August 27, 1918, the president appointed Ryan the Second Assistant Secretary of War and Director of the Army Air Service, created at this time by consolidating BAP and DMA. The appointment was a step toward representation of aeronautics at a higher level, but it also served to forestall creation of a separate department of aeronautics for which there was a great deal of congressional sentiment. Ryan was now Director of the Air Service as well as chairman of the civilian Aircraft Board.[109]

Airplane Engineering Division. On August 31, 1918, Nash created an Airplane Engineering Division within BAP by combining the Airplane Engineering Department and the Production Engineering Department. He also directed the Production Engineering Department to move to Dayton from Washington, D.C. The new division was given complete supervision over all engineering for the Bureau of Aircraft Production. Nash's directive also designated Lieutenant Colonel Vincent as chief of the Airplane Engineering Division, and Colonel Waldon associate chief.[110]

BAP defined the new Airplane Engineering Division as consisting of three main departments. Experimental Engineering at McCook Field under Major Marmon was responsible for all aircraft engineering up to the point of production, and for administering experimental work on contract with commercial organizations. Production Engineering under J.F. McClelland at the Air Service Building in Dayton was responsible for engineering on all aircraft in production. Business and Military, under Captain H.E. Blood, was also at the Air Service Building. This arrangement lasted until the armistice. To facilitate decisions on engineering and production and to coordinate the work of engineering and production with the Technical Section, Nash arranged for joint conferences to be held in Dayton between himself and the three chiefs.[111]

Just prior to the signing of the armistice, the Airplane Engineering Division included 58 officers, 385 enlisted, and 1,915 civilian personnel, making a total of 2,358 people. By January 1, 1919, McCook Field personnel declined to 1,474, including 56 officers, 322 enlisted, and 1,096 civilians.[112] At the time of the armistice, the Air Service had a total of 195,024 men (20,568 officers and 174,456 enlisted and civilians.) During the war, 16,952 aircraft had been delivered. Of this number, the U.S. produced 11,754 aircraft and the allies, mostly France, produced 5,198. The U.S. produced more than 1,000 balloons, including almost 650 observation types, and 32,420 aviation engines. At the armistice, the Air Service had 3,538 aircraft in Europe and 4,865 based in the U.S. In addition, Navy and Marine aviation personnel numbered 6,998 officers and 32,882 men with 2,107 aircraft (including 1,172 flying boats.) However, no aircraft entirely designed in the U.S. reached Europe in time to affect the war. Appropriations for the aircraft program exceeded $1.69 billion, a figure reduced by April 1921 to $598 million by revocation of contracts, cancellations, and salvage.[113]

On November 25, 1918, Colonel Thurman H. Bane succeeded Colonels Vincent and Waldon to command both the Airplane Engineering Division and McCook Field. After the armistice, Assistant Secretary of War Ryan resigned, effective November 27, 1918, leaving vacant four positions: Chairman of the Aircraft Board, Director of the Air Service, Director of the Bureau of Aircraft Production, and Director of the Spruce Production Corporation (which Congress had created to secure wood for the production of aircraft.)[114]

Secretary of War Baker wanted the Bureau of Aircraft Production brought under military control. With the concurrence of the Secretary of the Navy, he obtained an executive order from President Wilson on March 19, 1919. The order dissolved the Aircraft Board, placed the Bureau of Aircraft Production under the control of the Director of the Air Service, and vested the powers of the Director of Aircraft Production in the office itself rather than in the person holding it. Thus, by the spring of 1919, following months of change and stress, the Air Service achieved a measure of stability.[115]

Meanwhile, the Air Service appointed a new chief, Major General Charles T. Menoher, on December 23, 1918. He previously commanded the Rainbow Division in France. In the spring of 1919, General Menoher, who lacked aviation experience, organized the Air Service into four divisions: Information, Training and Operations, and Administration. The Supply Division, headed by Colonel William E. Gillmore, assumed the functions of the Bureau of Aircraft Production. After the reorganization in March, the Bureau of Aircraft Production and the Division of Military Aeronautics continued to exist only nominally while contracts and business obligations were closed out.[116]

Even in the spring of 1919, the Air Service was still considered temporary, and the organization was uncertain about its future. In July, however, the General Staff authorized funds for operating through June 30, 1920, and an Air Service of 1,000 officers and 11,000 enlisted. This plan of organization removed some of the uncertainty surrounding the Air Service. With Army appropriations for fiscal year 1920 cut by Congress, Secretary Baker issued an order discharging all temporary aviation officers by September 30, 1919. The plan provided that discharged officers could be retained as civilian employees.[117]

Technical and Engineering Divisions. Meanwhile, on January 1, 1919, a Technical Division was created at McCook Field by joining the Airplane Engineering Department (BAP), the Technical Section (DMA), and the Testing Squadron of Wilbur Wright Field. The Technical Division immediately changed its name to the Engineering Division on March 13, 1919, a name that remained in effect at McCook Field until establishment of the Air Corps on July 2, 1926. In May 1919, Langley Field's aircraft experimental activities transferred to McCook Field, making the Engineering Division the Air Service's center for development, testing, and procurement of aircraft, engines, and equipment. The Engineering Division reported through the Office of

Engineering Division in Washington, D.C. to a small Supply Group headed by Colonel William E. Gillmore. The group brought together all of the major elements of materiel logistics including research and development, procurement, supply, and maintenance.[118]

The short-lived Technical Division consisted of five departments: Airplane, Armament, Power Plant, Equipment, and Material. The Technical Division reported to two organizations since the Division of Military Aeronautics and the Bureau of Aircraft Production still existed in name. Colonel Bane continued as chief of the Technical Division and then of the Engineering Division until he retired in poor health on December 15, 1922.[119]

After establishment of the Engineering Division, the five departments of the former Technical Division reduced to three: a Military and Business Department; a Service Department; and an Engineering Department. The Engineering Department included five sections: Airplane, Power Plant, Equipment, Material, and Armament. The functions of the Engineering Division were similar to those of the organizations preceeding it at McCook Field. The Engineering Division handled all experimental and developmental work of a technical nature. It also performed all engineering changes, performance tests, and the collection of technical information, as well as the preparation, issuance, and safekeeping of all drawings, records and specifications. Thus, the Engineering Division continued the development, testing, and procurement of aircraft, engines, and equipment, but with limited funds and personnel. Much of the work was limited to improving items of war stock, particularly the Liberty engine.[120]

The Engineering Division of the Air Service continued until 1926. Organizational changes instituted by Major General Mason M. Patrick after he succeeded General Menoher as chief of the Air Service on October 5, 1921 (Patrick served until December 13, 1927) eliminated the Engineering Division's three departments. General Patrick restructured the division's functions into eight sections. The Power Plant Section continued as a major element of the Engineering Division.[121]

McCook Field at War

Technical Facilities. At the time of the armistice, McCook Field contained 254.38 acres of land, all graded and drained and the flying portion heavily rolled and sodded. To allow flying in bad weather, a macadamized takeoff area 100 feet wide and 1,340 feet long had been constructed on the field. By the armistice, 47 buildings had been constructed having a total floor space of approximately 371,914 square feet. Most of the buildings were steam heated from a central heating plant and an auxiliary plant located near one end of the field. Included among the buildings were administration, engineering, assembly, main hangar, engine assembly and dynamometer laboratory, propeller testing laboratory, research laboratory, propeller storage, exhibition hangar, storage hangar, lumber storage, central heating plant, emergency lighting plant, stock and receiving, gun range, machine shop, hospital, barracks, mess hall, cafeteria, post exchange, fire house, and garage.[122]

Most of the buildings were temporary structures of frame construction requiring attention to reduce the danger of fire. Modern fire equipment was installed throughout, a trained fire department was on duty at all times, and inspections of the entire field were made frequently. A stand-by fire protection system for use if the Dayton water supply was cut off included steam-driven fire pumps installed at the central heating plant and water supplied from wells drilled on the premises to a 55,000-gallon water tank raised on a 125-foot steel tower. The total cost of improvements at McCook Field, including all grading, driveways, service lines, sewers, buildings, machinery, and equipment exceeded $2.35 million.[123]

The engineering building was a two-story structure 60 feet wide by 600 feet long. The first floor contained the wood shop, machine shop, unit assembly and inspection, while the second floor contained the engineering offices and main drafting room. The shops, manned by skilled workers, were fully equipped to construct complete experimental aircraft, engines, propellers, and accessories.[124]

A steel and concrete reinforced building 66 feet by 260 feet housed the engine assembly and testing department. The dynamometer laboratory occupied a floor space of 66 feet by 120 feet; the engine assembly room, organization offices, stock room, wash rooms, and miscellaneous space took up the remainder. The dynamometer laboratory was divided into three testing rooms and each room contained two electric dynamometers. These dynamometers were installed end-to-end and aligned on the same center line so that they could be coupled and operated in tandem. This arrangement gave a maximum capacity of 800 horsepower when two 400-horsepower dynamometers were linked. Two test rooms had two 400-horsepower dynamometers each. The third room housed one 300- and one 100-horsepower dynamometer (used for testing the experimental single-cylinder Liberty engine.)[125]

Observers considered the dynamometer laboratory at McCook Field one of the finest of its kind. The building contained almost entirely invisible piping and wiring, allowing each dynamometer and test block a comparatively clear space for working around a test engine. The resistance units for the dynamometers, though easily accessible, were located out of the way on a mezzanine floor above the stock room. An efficient ventilating system removed the heat from this set up and kept the temperature moderate in the test room. Gases from the engines during tests were exhausted through a special system of ducts under the floor.[126]

McCook Field also had a portable dynamometer laboratory for obtaining performance data at high altitudes. At that time, the best test results were obtained by actually operating engines at high altitude. Since complete tests could not be made in aircraft, engines were taken to Pike's Peak at 14,109 feet above sea level. This Colorado site had a stone road leading to the summit, making it easy to transport the equipment. Points along the way to the top allowed testing at heights of 6,000, 7,500, 10,000, 11,000, and 12,500 feet. For this purpose, McCook Field technicians built a complete single test stand laboratory on a truck. One person controlled the compact, simple equipment from the instrument board. The test engine was mounted on a roller-bearing cradle torque stand. A propeller mounted on the engine shaft absorbed the power developed by the engine..[127]

The propeller test laboratory at McCook Field offered excellent facilities for conducting various whirling tests, including propeller calibration (to measure thrust, torque, and blade distortion at various speeds), endurance, and destructive testing. Experimental work included trials of special splices and of arrangements of woods in various combinations. Propeller designs proven satisfactory in one wood were replicated in other woods and comparison tested. The equipment could accommodate propellers absorbing up to 1,000 horsepower and with a maximum diameter of 18 feet and a speed range up to 3,000 revolutions-per-minute.[128]

The propeller laboratory's dynamometer equipment consisted of four 250-horsepower electric dynamometers aligned to operate in tandem. The dynamometers were entirely enclosed in a streamlined metal housing. All of the instruments and controls were installed within an explosion-proof observation room located below ground, directly under the propeller. The reinforced concrete roof of this room allowed observation of the tests through a series of narrow steel vertical slots cutting the plane of the propeller disc at right angles. An armored arch used as a traveling crane was moved into the plane of the propeller disc to contain any flying pieces in case the propeller broke.[129]

McCook Field technicians designed and built an instrument wind tunnel, modeling the tunnel after one in Orville Wright's laboratory. The preliminary design data was test records made on some 200 Venturi tubes of different sizes and shapes in the laboratories of carburetor companies. The tunnel was constructed using 2,100 pieces of high-quality walnut taken from the trimmings of propeller laminations. Each three-fourth inch length was made up of six segments glued and nailed in circular shape before truing in a lathe. A highly polished enamel surface lined the interior of the tunnel, while the exterior had a natural finish.[130]

The tunnel, almost 22 feet long, had a 14-inch diameter choke and an 80-inch diameter near the exit. The exit itself was 56 inches in diameter with a centrally located hemisphere 40 inches in diameter that acted as a shroud. The tunnel, mounted on a wood base 21 feet long and 6 feet wide, was cushioned by half-inch felt strips. Air was drawn through the tunnel by a 24-blade fan 60 inches in diameter. The fan was driven through a jack-shaft mounted on ball bearings and flexibly connected to a Sprague electric dynamometer. The dynamometer, rated at 200 to 300 horsepower when acting as a motor, drove the fan at a maximum speed of 2,000 revolutions-per-minute, producing an air speed of 453 miles-per-hour at the tunnel's choke. The control board allowed the operator to obtain any speed from 10 to 450 miles-per-hour. The speed of the air flowing through the choke was measured by built-in impact and static tubes registered on a differential water gauge.[131]

Air speed instruments tested or calibrated in the wind tunnel were mounted on a slide and placed into the choke through the plate glass observation door at the side. Facilities provided smoke pictures of the air flow through the choke so scientists could study the reaction of various shapes throughout a wide speed range.[132]

Wartime Accomplishments. During the 1920s, McCook Field was the single most influential agency in American air power, an effective center for aviation research and development. This leadership began to take shape during the Great War when McCook Field became the center of aircraft development and engineering activities for the Air Service. Even though McCook Field built several experimental combat aircraft, the armistice largely cut short their production. Most of the airplane work at McCook Field focused on modifying European models to adapt them to American production, a task of production engineering rather than experimental engineering. In general, McCook's activities began too late to affect combat operations.[133]

After establishment of the Airplane Engineering Department in October 1917, engineers pushed design of the USAC-1 (USA Combat 1), a two-seat fighter modeled on the British Bristol fighter. The USAC-1, purely an experimental design, was to take the high-compression, straight, spur-geared Liberty 12 engine, also being designed at the time. Another experimental model, the USAC-2, was a streamlined version of this aircraft. Although not markedly advanced, these aircraft set the precedent for McCook's engineering tradition.[134]

Engineers also redesigned the DeHavilland DH-9 aircraft, designated the USD-9 and the USD-9A in McCook Field versions, to carry the high-compression Liberty 12 engine. Three major McCook Field innovations were added to the aircraft: the Nelson gun control system, the Loomis cooling system, and the McCook Field gasoline system. This model added fuel capacity and a control that operated off the camshaft drive of the Liberty engine to time firing through the propeller disc. Although the USD-9A model was ordered into production, success at the front in 1918 caused cancellation of the contract with Curtiss.[135]

Another redesign project involved the Bristol F-2B fighter to take an Hispano-Suiza 8-cylinder, 300-horsepower engine. The McCook Field version was designated the USB-1. Two experimental versions of this design, designated USXB-1 and USXB-2, were made. The USXB-2 was built around the 8-cylinder Liberty engine. This aircraft was also set aside after wartime success.[136] Similarly, the Pomilio FVL-8, a fighter using the 290-horsepower Liberty V-8, and the Pomilio VL-12, a bomber using the Liberty V-12, were cut short by success on the battlefield. These aircraft were the product of Octavio Pomilio, an Italian designer on loan to McCook by his government.[137]

McCook Field also supported a great deal of experimental work performed by outside engineering organizations. Engineering administration of these contracts was handled through McCook's experimental engineering organization and business administration by the field's business headquarters. Numerous firms worked on experimental engines, special ignition and carburetion problems, propeller reduction gears, radiators, propellers of steel and micarta duck construction, and different materials for application to aircraft. Experimental work was supported for the development of turbo-superchargers, turbines with interconnected compressors driven by exhaust gas, so engines could continue to operate effectively in the decreased atmospheric density at high altitudes. Contractual services were used to develop several highly regarded aircraft such as the LePere two-seat fighter and observation plane, the Martin bomber, the Vought advanced trainer, and the Loening monoplane two-seat fighter.[138]

The Packard Motor Car Company manufactured the LePere LUSAC-11, designed by French Army Captain Georges LePere on loan to McCook Field by his government. The LUSAC-11, a two-seat fighter and observation aircraft similar in appearance to the Bristol fighter, used one Liberty 12-cylinder engine. The first aircraft was manufactured in the summer of 1918, and trial flights began in August. Ultimately, 27 aircraft were produced. Although the aircraft went overseas, it did not fly against the enemy.[139]

Glenn L. Martin's bomber, the GMB-1, was a large biplane equipped with two Liberty 12-cylinder engines. The bomber made its first flight on August 17, 1918. A contract let in October for 50 aircraft, but at war's end the number reduced to four. Nonetheless, the GMB-1 established a tradition of big U.S. bombers and influenced aviation policy. Lewis and Vought Corporation's VE-7 advanced trainer, a two-seat biplane using an Hispano-Suiza 180-horsepower engine, was tested at McCook Field. The follow-on version, designated VE-8, was found unsatisfactory during tests at the field.[140]

Grover C. Loening's model M-8 monoplane two-seat fighter, a design started from scratch, used a 300-horsepower Hispano-Suiza engine. The aircraft had a top speed of 146 miles per hour and reached an altitude of 26,000 feet. Loening's aircraft would have gone into production had the war continued into 1919, but only two were purchased before the war ended.[141]

McCook Field design engineers accomplished many projects related to improving aircraft operations, including standardized drawings of the Liberty 8-cylinder engine; design of the 1-cylinder and 2-cylinder Liberty test engines; design of the Liberty epicyclic reduction gear; the Nelson gun control for synchronizing machine gun fire through the propeller disc; propeller designs for various aircraft; standardization of utility parts; and preparation of the Standard Book for the work done on airplanes and engines by American manufacturers and government departments.[142]

McCook Field's experimental factory performed the shop work, including practically all phases of aircraft and engine construction. These projects covered propellers, accessories, and armament. The shop made the small Liberty test engines and the 24-cylinder Liberty engine. It also developed the Nelson gun control, fabricated 50 samples, and sent them to different manufacturers for production.[143]

The Engine Assembly and Testing Department was able to test all types of aviation engines and accessories, including radiators, gasoline systems, ignition systems, and heating and lighting systems. The department's extensive research and experimental work contributed to the effectiveness of the Liberty engine. The department tested experimental engines and accessories, and production engines sent to McCook for calibration or for installation in experimental aircraft built at the field or by manufacturers. It also did routine overhauling and testing of engines in use at the field. The department tested several different gasoline systems in developing the McCook Field system used on the USD-9A. This system consisted of a main supply tank that fed the carburetors by means of two automatic relief gasoline force pumps of the vane type driven from the engine crankshaft by flexible shafting and included an auxiliary gravity tank located in the center panel of the upper wing. The pump arrangement, which provided positive feed under all conditions, did not use air pressure in the fuel system in order to minimize the danger of fire from fuel leaks. The system provided a positive gravity feed at starting.[144]

The Mechanical Research Department, directed by Allen Loomis who came from the automotive industry, developed an efficient cooling system for the Liberty engine. The department conducted bench and flight tests on different types and designs of radiators and cooling systems before devising a system satisfactory for use in combat. This Loomis cooling system had two innovative features. The first feature was an expansion tank that surrounded the core honeycomb cooling element and formed an integral part of the nose radiator. The second feature was an injector in the water connection between the radiator and the circulating pump; this injector drew water through a nozzle outlet from the bottom of the expansion tank and injected it into the circulating system, thereby keeping a constant volume of water circulating. Pressure circulated the water faster, increasing cooling efficiency, and minimizing water loss.[145]

The Propeller Department first organized in Washington, D.C. as part of the Plane Design Section by F.W. Caldwell who transferred from Curtiss. The department developed aircraft propellers for the Air Service and the Navy. The department also did considerable experimental and research work on propellers and related problems, designing and testing them both in flight and to destruction in whirling tests. Designs of variable pitch propellers were tested to destruction and flight tested. Propeller thrust and torque meters were used in flight testing. Various woods and combinations of woods were tested to determine their suitability for use in propellers. The department experimented with methods of tipping propeller blades with various materials to protect against abrasions, including copper, brass, pigskin, linen, and other substances. Aluminum leaf coating was tested as protection against moisture. Micarta duck construction was applied to propellers, and steel propellers were fabricated. The department developed the McCook Field propeller testing laboratory. It also developed a standard propeller storage building equipped with automatic conditioning machinery to regulate temperature and humidity.[146]

A.L. Nelson, an engineer who had made a study of aircraft armament in Europe in order to solve the problem of synchronizing the firing of machine guns through the propeller disc. The department developed a mechanical system, officially designated the Nelson gun control, together with single-shot trigger mechanisms for several types of aircraft guns. The Nelson system, developed primarily to use with 4-bladed propellers, differed from the Constantinesco system which used pressure waves to transmit timing impulses from the engine to the gun through a tube containing a column of oil under pressure. The Nelson control eliminated the lag by transmitting the timing impulses through a positive mechanical connection between the engine and the gun. This new system proved more accurate with 4-bladed propellers and eliminated stray shots when used with the single-shot mechanism. The department also developed an ammunition testing machine.[147]

The Chemical and Physical Laboratories at McCook conducted tests on gasoline, lubricating oils, metals, and alloys. These laboratories also performed special experiments, including a complete series of spark plug tests.[148]

Lieutenant Alexander Klemin, a student of Professor Jerome C. Hunsaker and his successor at MIT, headed the Aeronautical Research Department. The department conducted the first aircraft performance tests at McCook Field. A special experimental aircraft assigned to this department was converted into a flying aerodynamic laboratory. In addition, the department conducted sand load tests of aircraft structures to simulate forces encountered in flight. MIT also conducted model testing of experimental aircraft, various wings, and propeller sections for McCook.[149]

The flight test organization at McCook Field, which had very few accidents and no fatalities during the war, regarded its mechanics and test pilots as the best available. They performed all kinds of flying tests at the field, including tests on experimental and production aircraft, engines, propellers, radiators, carburetors, ignition systems, standard and special instruments, parachutes, and accessories. Instruments tested included air speed indicators, oil pressure gauges, tachometers, and thermometers. The pilots calibrated instruments, did climbing and speed tests, and tested propellers and fuel consumption at various speeds and altitudes.[150]

McCook Field's Technical Publications and Library Department provided technical data and information for the engineers, and kept current the aeronautical reference library which contained hundreds of books, pamphlets, and magazines. The department translated and reviewed foreign technical reports and publications. Thousands of reports and documents, both foreign and domestic, were obtained through various agencies. After classifying and cross-indexing the data and reports, the department distributed them. The department regularly published and distributed the results of experimental and research work conducted both at McCook and by outside experimental engineering organizations, issuing weekly progress reports on the activities of experimental engineering, and publishing a monthly bulletin.[151]

Production engineering was another McCook Field responsibility. Thus, a Materials Department performed materials engineering and prepared specifications for aircraft production. An Ordnance Department dealt with the installation of armament and armament accessories on all aircraft, and contributed to the installation of the Nelson gun control on various aircraft. The Electrical and Radio Department worked on original designs of electrical heating and lighting devices in cooperation with other departments and manufacturers, and supported installation of radios in aircraft. A Planes Department with engineers in charge of specific aircraft monitored their production progress at the manufacturing plants. An Engineering Records Department maintained records and files on all government aircraft production contracts, and developed instruction books -- including all changes in design, materials, and specifications -- in cooperation with the Technical Section of DMA. The Technical Section also distributed technical data to the field, including technical orders, engineering orders, and handbooks on airplanes, engines, and accessories.[152]

The Instrument Department generally did no experimental work but did perform the actual flight testing of instruments installed on aircraft. Thus, the department was responsible for various oil and gasoline gauges, tachometers, air speed indicators, radiator and oil thermometers, and other aircraft accessories. The department worked on leak-proof gas tanks, examining and testing various gasoline systems in this connection. Hundreds of referrals came from outside sources, and the department commented on and criticized designs for new instruments or modifications of standard ones. Because some of the department's efforts overlapped work being done by the Science and Research Department, their areas of responsibility had to be defined late in 1918.[153]

After the fighting stopped in November 1918, many people left McCook in the months that followed. The Great War had left a legacy of an aviation engineering laboratory in Dayton. The nucleus that remained after the war built that legacy into a tradition of excellence.

CHAPTER 2 REFERENCES

Most of the sources cited in Chapter I are also useful for this chapter. In addition, two studies central to understanding U.S. aviation progress during the Great War are cited by everyone writing about this period. The first is a massive compilation in eight volumes edited by Russell M. McFarland, History of the Bureau of Aircraft Production (n.d., c. 1919.) The copy of McFarland cited here, filed in the archives of the Air Force Materiel Command at Wright-Patterson AFB, was reproduced in September 1951 with slight editorial changes. McFarland's precious work, sometimes lacking attribution, incorporates tremendous detail on every aspect of BAP's involvement. McFarland provides no index, but he has a detailed table of contents. The second study is a brief illustrated volume by Captain H.H. Blee, History of Organization and Activities of Airplane Engineering Division, Bureau of Aircraft Production (15 Aug. 1919), filed in the Base's History Office at WPAFB. Blee's work, incorporated by McFarland into his mammoth history without attribution, also has no index.

Excellent discussions of McCook Field -- and a good deal more -- are in the illustrated history by Lois E. Walker, Shelby E. Wickam, et al., From Huffman Prairie to the Moon: The History of Wright-Patterson Air Force Base (Washington: Government Printing Office, 1986.) Several articles covering McCook Field, focusing largely on the 1920s, are cited in the reference section of the next chapter. Those articles also recount McCook's contributions during WWI, but their discussions follow Blee and McFarland.

Several works that deal with the U.S. war effort date from the immediate postwar years: Colonel G.W. Mixter and Navy Lieutenant H.H. Emmons, United States Army Aircraft Production Facts (Washington: Government Printing Office, 1919); Arthur Sweetser, The American Air Service: Problems of War and Reconstruction (New York: D. Appleton & Co., 1919); and Colonel Edgar S. Gorrell, The Measure of America's World War Aeronautical Effort (Burlington, VT: Lane Press, Inc., 1940.) Data on the wide range of Army armament and equipment, including aviation production, was provided by Benedict Crowell, America's Munitions, 1917-1918 (Washington: Government Printing Office, 1919); Crowell provides numerous photographs. Although these works tend to be apologies about the U.S. production program in the Great War, they were necessary in the wake of the Hughes investigation. The statistics sometimes require close reading to reconcile, but the numbers are indicative. Only Sweetser has an index, but two of the others are brief. For countervailing views of U.S. accomplishments in the Great War, see General Henry H. Arnold, Global Mission (New York: Harper & Brothers, Publishers, 1949); and Grover C. Loening, Takeoff Into Greatness: How American Aviation Grew So Big So Fast (New York: G.P. Putnam's Sons, 1968.) Loening was especially bitter about the "Detroit gang."

All of the works cited refer to organizational highlights to provide perspective, but four other works supply detailed analysis. Royal D. Frey, Evolution of Maintenance Engineering, 1907-1920 (WPAFB: Air Materiel Command, July 1960), covers developments through the Great War with a highly readable text and several volumes of documentation. Edward O. Purtee, History of the Army Air Service, 1907-1926 (WPAFB: Air Materiel Command, May 1948), and Charles C. Mooney & Martha E. Layman, Organization of Military Aeronautics, 1907-1935 (Washington: AAF Historical Office, Dec. 1944), extend the discussion of organizational progress to the formation of the Army Air Corps. Martin P. Claussen's Materiel Research and Development in the Army Air Arm, 1914-1945 (Washington: AAF Historical Office, Nov. 1946), takes the discussion through WWII.

Development and production of the Liberty engine is universally conceded to be the greatest U.S. technical contribution to WWI. The best summary of its history is provided by Lieutenant Colonel Philip S. Dickey III, The Liberty Engine, 1918-1942 (Washington: Smithsonian Institution Press, 1968.) More general information on the Liberty and other aircraft piston engines is in C. Fayette Taylor's Aircraft Propulsion: A

Review of the Evolution of Aircraft Piston Engines (Washington: Smithsonian Institution Press, 1971.) Deeds' role in the Liberty's development from his viewpoint is told by Isaac F. Marcosson, Colonel Deeds: Industrial Builder (New York: Dood, Mean & Co., 1947.)

[1] Frey, 188. According to Frey (page 80), the U.S. Army's regulation requiring officers to wear spurs while flying was active for a year into World War I.

[2] Sweetser, 251-252. Claussen, 17.

[3] Hennessy, 155-156. Purtee, 33-35. Mooney & Layman, 27. McFarland, 568-569, 583-585, 614-615, 692-693. Sweetser, 47-50. Holley, 39-40.

In April 1917, NACA also recommended that the Council of National Defense establish a Joint Army and Navy Technical Board. The council did so, forming a board composed mostly of experienced aviators. That joint board considered matters such as available and required types of aircraft, and aircraft designs and specifications. The joint board also made recommendations to the Aircraft Production Board which, in turn, recommended placing contracts with the various manufacturers. The contracts were then initiated by the financial unit of the Signal Corps. Foulois, 150.

[4] The Aircraft Board held its first meeting on November 6, 1917 and the Aircraft Production Board held its last meeting on November 3, 1917. The last meeting of the Aircraft Board was held on November 29, 1918.

The Aircraft Board was created by Congress on October 1, 1917 to succeed the Aircraft Production Board as an advisory body, thereby giving a legal basis to the board's activities. The congressional act also transferred control from the Council of National Defense to the Secretaries of War and Navy. The change occurred through the influence of Coffin, who served as chairman of both boards, because he believed that establishing a separate department of aeronautics would be premature. The Aircraft Board was meant to function as a clearinghouse between the Signal Corps and other agencies, including the General Staff, the Army, Navy, and industry in matters pertaining to aviation. Congress empowered the Aircraft Board to supervise and direct, in accordance with the requirements of the respective departments, the purchase, production, and manufacture of aircraft, engines, and all ordnance and accessories. Procurements were to be made through the regular purchasing departments of the Army and Navy, making the main function of the board advisory. Limitations imposed by the Army and Navy denied it the right of direct communications with manufacturing plants. Purtee, 62-63. Frey, 117. Sweetser, 93-94.

[5] Mooney & Layman, 27-28. McFarland, 591-592, 603, 615-617, 629, 694. Frey, 117. Hennessy, 196.

[6] After its creation, the Aviation Section continued to be referred to as the Aeronautical Division, adding to the organizational confusion of this period. Several other names were also used from time to time until the formation of the Air Service. Arnold, 56-57.

[7] Purtee, 34. McFarland, 21, 261, 620, 635-654.

[8] Major Henry Souther was born on September 11, 1965 and died on August 15, 1917 after an operation. He was survived by a wife and two daughters. At the time of his death, Souther headedf the Aircraft Engineering Division of the Air Service. Souther received his training in engineering at MIT, graduating in 1887. He was an engineer for the Pennsylvania Steel Company for five years and then with the Pope Manufacturing Company for six years. Souther became one of the nation's leading authorities on materials, and opened a consulting engineering laboratory in Springfield, Massachusetts. He served as water commissioner in Hartford for eight years. Souther promoted standardization and served as chairman of the standards committee of the Society of Automobile Engineers. He was president of the SAE in 1910-1911. General Squier called Souther to Washington to be consulting engineer of the Aviation Section, Signal Corps, in April 1916. He was the first civilian to accept a commission in the Officers' Reserve Corps of the Aviation Section as a major. Souther organized the inspection department and conducted experimental testing. One of his duties was to act as director

of Langley Field. When he died, his work was largely connected with the experimental testing of machines and engines. "Major Henry Souther Dies," Aviation, III:2 (Aug.15, 1917), 111. For his assessment of the state-of-the-art of aviation engines before the U.S. entered WWI, see Henry Souther, "Development and Progress in Aviation Engines," Aviation, III:12 (Jan. 15, 1918), 833-835; he presented this paper in 1916.

9   Biographical information in the files of the Air Force Museum shows that Clark (Feb. 27, 1886-Jan. 30, 1948) was a singular individual. Born in Uniontown, Pennsylvania, he was graduated from the Naval Academy in 1907. After a world cruise with the battleship fleet, he was stationed in China. He served with the Coast Artillery, learned to fly, and transferred into the Signal Corps. He paid his own way through the graduate program at the Massachusetts Institute of Technology, studying with Jerome C. Hunsaker. After receiving his degree in aeronautical engineering from MIT in July 1915, Clark was immediately placed in charge of the Experimental and Repair Department at the Signal Corps Aviation School at North Island. When the U.S. entered World War I, Clark was regarded as the leading aircraft expert in the Army. He served on the Bolling mission to Europe in 1917.

Following service as the first commanding officer of McCook Field, Clark transferred to Washington as aircraft designer, apparently forced out by Jesse Vincent's maneuvering. After WWI, he returned to McCook Field in January 1919 and headed the Aircraft Section of the Engineering Division. Clark supported the design of various successful aircraft designed by the Engineering Division. He was court martialed in 1920. He was convicted by both the court martial and a civilian court of forty counts of bigamy and discharged on 15 November 1920 as a Class B officer, that is, "under less than honorable conditions."

Clark's discharge from the Army marked the beginning of a distinguished civilian career. After 1920, he was chief aeronautical engineer for the Dayton-Wright Division of General Motors Corporation. In 1922 and 1923, he also served as vice president of the Society of Automotive Engineers. In 1923, he helped organize the Consolidated Aircraft Corporation which took over the designs of the Dayton-Wright Company. Thereafter, Clark served as vice president and chief engineer for several other companies. In 1927, he was U.S. representative to the International Aviation Congress in Rome. He became president of Clark Aircraft Corporation. During World War II, he was consulting engineer for Hughes Aircraft Company.

By 1922, Clark designed many aircraft and supervised the designing of others. He also designed practically all U.S. wing sections to that time. The Clark "Y" airfoil came out in 1924 and was still in use in 1942. He also designed Clark "V" series airfoils, and U.S.A. airfoils 16 and 27. Many successful aircraft used the Clark airfoils, for example, the Spirit of St. Louis, the Douglas around-the-world aircraft, and Army and Navy pursuit and training planes. Clark received many patents. He published a book, Elements of Aviation, in 1928 and wrote many articles on aviation.

Clark also received credit for antedating Fred E. Weick's work with the NACA cowl of 1928. Experimental efforts to reduce the drag of radial engines began as early as 1918, but the need to provide adequate cooling and accessibility made designers cautious. Rotary engines had been cowled but their rotation made cooling them easier. Various efforts by designers were tested in wind tunnels and aircraft. The Dayton-Wright XPS-1 aircraft in 1922 had a cowl designed by Clark similar to those used on the Deperdussin rotary engine and like NACA's later design. The aircraft, which used a Lawrance J-1 radial engine, was not successful and its retractable undercarriage received more attention than the cowl. Moreover, Clark could not explain how the cowl worked, making it difficult to persuade others to use it. Nevertheless, Pratt and Whitney's chief engineer, G.J. Mead, used Clark as a consultant in 1927 and wanted to fit the cowling on the Corsair aircraft that Chance Vought was designing around Pratt and Whitney's first engine, the Wasp. Vought did not use it. Ronald E. Miller and David Sawers, The Technical Development of Modern Aviation (London: Routledge & Kegan Paul, 1968), 62-63, 248.

McCook Field's <u>Slipstream</u> publication featured Clark in December 1919 (pages 8, 31) and Christmas 1920 (page 13). See also, "Col. Clark Joins Dayton Wright," <u>Aviation</u>, IX:13 (Dec. 13, 1920), 413, 419.

[10]   Blee, 1. McFarland, 275, 1829. Frey, 54, 109.

[11]   NACA was located in the Munsey Building from its inception until late 1917. The Aircraft Board, which succeeded the Aircraft Production Board, also was in the Munsey Building until it moved in the summer of 1918 to Building D on 4th Street and Missouri Avenue.

[12]   Purtee, 35, 50. Frey, 109-110.

[13]   Purtee, 1-2. McFarland, 275, 628. Frey, 109.

[14]   Biographical records in the Air Force Museum and Dickey (pp. 11-12) indicate that Vincent (Feb. 10, 1880-April 20, 1962) was another remarkable individual. Born in Charleston, Arkansas, Vincent attended schools in Illinois and St. Louis, Missouri. He learned engineering through correspondence courses, and, at age 18, became a machinist and toolmaker. Between 1903 and 1910, he was superintendent of inventions for Burroughs Adding Machine Company in Detroit. From 1910-1912, he was chief engineer for Hudson Motor Car Company. Between 1912 and 1917, he was vice president of engineering for Packard Motor Car Company, gaining enough patents to fill a book.

In 1915, Packard began developing an aircraft engine, but this initial effort proved too heavy. In May 1917, Vincent contacted government officials about building a standardized engine. He and Hall began laying out the design for a new engine in a room in the Willard Hotel in Washington, D.C., on May 29. They completed the the design by May 31. Practically all the features of the new engine had already been proved by experimental work and manufacturing experience at Packard and Hall-Scott. Vincent remained the prime mover behind development and testing of the engine. Named the Liberty 12, it became America's most notable technological achievement in WWI. It proved superior to any other aircraft engine of its time and continued in use for more than a decade.

After completing work on the Liberty engine, Vincent accepted a commission as a major in the Signal Corps on August 15, 1917. He was instrumental in setting up the experimental engineering station at McCook Field in October 1917. On February 6, 1918, he was placed in full charge of the Airplane Engineering Department as a lieutenant colonel. Before the war ended, Hughes -- ignoring Vincent's contribution to the war effort -- accused Vincent of profiteering because of contracts awarded to Packard for production of the Liberty engine while he was still a stockholder. President Wilson granted a full pardon on December 3, 1918. Meanwhile, Vincent was honorably discharged on November 30, 1918 and recommended for a Distinguished Service Medal. However, the medal was not awarded because of the Hughes controversy.

After World War I, Vincent returned to Packard. But in 1940, with World War II underway in Europe, the government asked Packard to produce the Rolls-Royce Merlin engine which was being assembled by hand in England. Packard made Vincent responsible for re-engineering the British engine to provide for U.S. carburetors, propellers, propeller governors, vacuum and fuel pumps, generators, and other accessory items. Vincent led the project through to completion, producing a design superior to the original English engine. The U.S. and England received more than 55,500 Merlin engines from Packard for installation in the P-51, Spitfire, and Mosquito aircraft. Vincent retired from Packard in 1948, recognized by President Harry S. Truman for his contributions in WWII.

Vincent's brother, Charles H. Vincent, was also an automotive engineer who worked on aircraft engines. Charles, writing in 1971 at age 86, felt that his engineering contributions to the Liberty had not been properly recognized. He considered Colonel Philip S. Dickey's 1968 book on the Liberty engine accurate "on the whole," but "with some errors and omissions."

[15]   Blee, , 2. McFarland, 276, 2054, 2109, 2156-2157. Frey, 110-111.

[16]   Purtee, 59-60. Dickey, 56. Sweetser, 72-74, 90, 211, 231-232, 254-256, 341-347. Goldberg, 14.

In April 1917, NACA also recommended that the Council of National Defense establish a Joint Army and Navy Technical Board. The council did so, forming a board composed mostly of experienced aviators. That joint board considered matters such as available and required types of aircraft, and aircraft designs and specifications. The joint board also made recommendations to the Aircraft Production Board which, in turn, recommended placing contracts with the various manufacturers. The contracts were then initiated by the financial unit of the Signal Corps. Foulois, 150.

[17]   Blee, 3. McFarland, 276-277, 593, 621-627. Purtee, 58-59. Mooney & Layman, 28-29. Frey, 111. Sweetser, 93-94. Holley, 68.

[18]   Blee, 3-4. Purtee, 58-59, 73. McFarland, 24. Frey, 112.

After Deeds restructured the Equipment Division, the Signal Corps further reorganized. By September 1917, the Signal Corps included eight divisions in addition to the Equipment Division: Air (for recruiting and training), Radio, Photography, Construction (of flying fields), Science and Research, Langley Field, Administration, and Land. All of the divisions except Land involved aviation. Purtee, 59-60.

[19]   Purtee, 24-25, 27. McFarland, 383-389, 563-564, 567, 572, 577, 581. Frey, 62, 70-71. Claussen, 14-15.

The war ended before the first NACA wind tunnel could be completed and before its new engine dynamometer buildings became operational. Actually, NACA's wartime research activities were less prominent than its advisory work on other aeronautical matters — production planning, standardization, patents, and training.

[20]   McFarland, 390-394. Claussen, 15.

The following summary is based on McFarland, 392-393, 395-397, 399-402:

Albert Kahn, an architect from Detroit, prepared plans for Langley. His plan was approved by the Secretary of War in May 1917, and a contract for construction was made with the J.G. White Engineering Corporation. The Construction Division of the Signal Corps, created on May 21, 1917, assumed responsibility for the construction of Langley from the Quartermaster Corps. NACA's buildings were designed to conform with the general architectural plans prepared by Kahn.

Construction started at once, but many delays occurred because of difficulties obtaining materials and skilled labor. Soon after construction began, a highway was built to connect the field with Hampton, Virginia. A trolley line was constructed parallel to the highway and a branch line connected with the Chesapeake and Ohio railway. A bridge was constructed over a branch of the Back River to bring the highway and railroad into the field.

The Equipment Division spent considerable money on the field for wind tunnel and other improvements and used the field for work on instruments, ordnance, bombs, and related equipment. On August 22, 1918, the Bureau of Aircraft Production formed an Experimental Engineering Department at Langley Field.

In July 1918, the Supply Section of Division of Military Aeronautics began taking over all construction work at Langley Field. White Corporation began withdrawing from Langley Field in early August 1918. Construction proceeded under the Supply Section until the end of November when the work was discontinued.

In January 1919, Langley Field had 23 permanent structures, mostly family residences, completed. Sixteen other permanent structures were partially completed, again mostly residences. Construction to that date cost $8 million, and an additional $7 million was needed to complete the architectural plan and other work. At the signing of the armistice, the Bureau of Aircraft Production had some R&D work underway at Langley Field. The work fell into four categories, as described by the Science and Research Department: photographic study of bomb trajectories; development of a stabilized bombsight; development of instruments and methods for aircraft navigation for extended flights; and aircraft signaling projectors and night recognition lights. Total military personnel at Langley Field at the signing of the armistice was reported as 187 officers, 278 student officers, 18 cadets, 2,028 enlisted white men, and 2,268 enlisted black men.

[21] McFarland, 2063.

[22] McFarland, 2066.

[23] McFarland, 2066.

[24] McFarland, 2122-2123.

[25] McFarland, 2123.

[26] McFarland, 2110-2111.

[27] McFarland, 2111.

[28] McFarland, 2111-2112.

[29] Blee, 78-79. McFarland, 355-356, 1132-1133. Purtee, 63-64.

[30] Blee, 80. McFarland, 357. Frey, 112-113.

The summary below is based on Crouch, Bishop's Boys, 468-471, and Walker-Wickam, 23, 88, 183:

The Dayton-Wright Airplane Company was organized in March 1917 by Deeds and his business associates after the Wright-Martin Company moved from Dayton to New Jersey. At that time, Deeds, who had a national reputation as an industrialist, was serving as a member of the U.S. Munitions Standards Board in Washington. Deeds' friends included Orville Wright, Charles F. Kettering, and the Harold Talbotts, father and son. Deeds and Kettering earned reputations while working under John H. Patterson at the National Cash Register Company. They formed a company of their own in 1914, the Dayton Engineering Laboratories (Delco), to produce an automobile self-starter developed by Kettering in Deeds' barn. They invested the profits from that venture in a second firm, the Dayton Metal Products Company, which they founded with the Talbotts, local building contractors. The Dayton Airplane Company was their next joint project.

Except for Orville Wright, who signed on as a consulting engineer, Deeds and his associates knew little about airplanes but they believed that their experience in production would help them with military contracts. Coffin, head of the Aircraft Production Board, was a friend of Deeds and Kettering. With American entry into WWI, Deeds was commissioned a colonel in the Signal Corps Reserve and placed in charge of procurement for the Aircraft Production Board as chief of the Equipment Division. He divested himself of financial interest in the Dayton-Wright Company in April, then awarded the company two contracts for production of 4,400 aircraft, 4000 DeHavilland DH-4 two-place biplanes and 400 J-1 trainers designed by an American firm, Standard Aircraft. By the end of 1918, the Dayton-Wright Airplane Company manufactured 3,106 of the total 4,500 DH-4s built in the U.S. and 400 SJ-1 trainers. Dayton-Wright found the task of building the DH-4s was more difficult than company officials had supposed. The first problem was to prepare a complete set of drawings, including the basic design changes required to accommodate the Liberty engine rather than the Rolls-Royce. Orville was heavily involved in these preparations, but he was more interested in another project -- Kettering's Bug, an unmanned flying bomb powered by a 4-cylinder engine that was still in development at the armistice.

Dayton-Wright operated in buildings constructed at South Field, in Moraine City, south of Dayton, on land owned by Deeds. In 1916, Deeds established a private flying field at his south Dayton estate, Moraine Farm. He equipped South Field with a hangar and a research laboratory, and permitted its use later as a testing ground for airplanes manufactured by the Dayton-Wright Airplane Company. In 1917, Deeds also purchased options on land adjoining Triangle Park which became known as North Field.

Dayton-Wright survived the scandal of Deeds' favoritism toward his friends. After the war, the firm produced a series of aircraft. General Motors bought the company in 1919 and kept it going until June 1, 1923 when GM got out of the aircraft business.

[31]   McFarland, 2112.

[32]   Blee , 80-81. McFarland, 357-358.

[33]   Blee, 81-83. McFarland, 358-359.

According to Vincent, *"designing of the buildings was carried out simultaneously with the construction work. In other words, there was no time to have the architect draw up plans and specifications before starting actual construction work."* McFarland, 2115.

[34]   Blee, 4-5, 41. McFarland, 278. Purtee, 60, 99-100. Frey, 112.

[35]   Blee, 39. Purtee, 64, and Appendix C. McFarland, 310-311. Frey, 125.

[36]   McFarland, 2113.

[37]   McFarland, 2113-2115.

[38]   Blee, 5. McFarland, 278-279, 2116-2117. Frey, 113. Purtee, Appendix C. Sweetser, 199-203, 244. Claussen, 18.

The misadventures of trying to produce the Bristol fighter while redesigning the aircraft to adapt the larger, heavier Liberty engine (and the Spad's similar fate) are discussed by Holley, 124-126.

[39]   McFarland, 2126-2128.

[40] Blee, 6. Purtee, 65. McFarland, 279-280, 394. Frey, 113, 125.

[41] Blee, 83-84. McFarland, 360.

[42] Blee, 6-8, 39. McFarland, 29, 280-281, 311. Frey, 114, 125. Purtee, Appendix C.

[43] Blee, 8-9, 40. Purtee, 100, and Appendix C. McFarland, 281-283, 311-312, 2147. Frey, 114, 125.

[44] Deeds told Vincent that he did not want to take him away from experimental work, and in fact placed him in charge of work on the Bugatti engine. Vincent acceded to Deeds' decision and focused on the Bugatti work. But in late January 1918, that work was moved to New York and Deeds made Vincent chief engineer of the Airplane Engineering Department. McFarland, 1884-1888, 2130-2133.

[45] Blee, 11-12. Purtee, 100-101, 106. McFarland, 30, 284-287. Frey, 115.

[46] Blee, 9-10, 39-40. McFarland, 283-284. Frey, 114.

[47] Blee, 13-15. McFarland, 286-288. Frey, 116-117.

[48] Dickey, 81. Frey, 136.

[49] Dickey, 81. McFarland, 1899, 2055. Claussen, 18-19.

Grover Loening, who put the inverted Liberty into widespread use with his post-WWI amphibian, was skeptical about the generally accepted accout of the origin of the Liberty engine, calling it a "yarn" that "maligned" the Liberty engine at first "because of the ridiculous publicity story of its birth that the auto people, who fathered it, were stupid enough to broadcast." Loening called the original Vincent-Hall engine "an inferior engine with eight cylinders that showed up so badly on the test stand that the War Aircraft Production Board had to form a committee of the best American experts to help rescue it from failure." Claiming that other automotive engineers "literally redesigned the motor" by making it into a 12-cylinder and by changing its oiling system, Loening granted that the engine "was far ahead of anything the Europeans were capable of." Grover Loening, Takeoff Into Greatness (New York: G.P. Putnam's Sons, 1968), 135-136.

[50] Dickey, ix. Sweetser, 184. Goldberg, 16.

[51] Bilstein, 34.

[52] Bilstein, 36. McFarland, 1828.

Sweetser, 148-167, 248-251, discusses at length the shortage of raw materials, especially spruce for construction of aircraft. McFarland's eight volumes provide detailed discussions of material shortages, among many other things, including spruce, hardwood, fabrics, dope, metals, and lubricants.

[53] Dickey, 1, 4, 7.

[54] Dickey, 3. Sweetser, 51, 168-169. McFarland, 1826-1827, 1843-1845.

[55] Gorrell, 1. Sweetser, 66, 71, 231. Goldberg, 14. Holley, 41-42. Foulois, 143-147.

General Henry H. (Hap) Arnold pointed out that Ribot's telegram was instigated by Major Billy Mitchell who arrived in Europe a week before the U.S. declared war. Mitchell already was advocating the bombing of Germany as the quickest way to break the stalemate on the front lines. The enormity of the problem is suggested by other statistics given by Arnold: the total strength of the American air arm in May 1917 was 52 officers and 1,100 men, plus some 200 civilian mechanics. Altogether, the U.S. air arm had 55 aircraft, "51 obsolete, 4 obsolescent, and not one of them a combat type." Actually, wrote Arnold, "it was worse than that. Statistics aside, we had not air power at all." Arnold, 49-50.

[56] Sweetser, 66-68, 72-74, 90, 211, 231-232, 254-256, 341-347. Goldberg, 14. Dickey, 3. Gorrell, 1, 7-8. Holley, 41-46. Foulois, 143-147.

[57] Dickey, 6. Sweetser, 64-65. Foulois, 148-153.

[58] Dickey, 7-8.

[59] Dickey, 7.

[60] Dickey, 7-8.

[61] Dickey, 8.

[62] Dickey, 8-9. Goldberg, 16-17.

[63] Dickey, 8-9.

[64] Hall was born in San Jose, California, on April 8, 1882. After correspondence classes and night courses, he became a steam engineer at the age of 16 with the I.L. Benton Machine Works in San Francisco. Within four years, he became part owner of the company, gaining experience in the design of marine, hoisting, and gasoline engines. In 1903, the company began building automobile engines. By 1905, Hall was working with the Heine-Velox Company producing automobiles. But both companies were wiped out in the earthquake of 1906.

In 1910, after a period of building an automobile of his own design, Hall joined with Bert C. Scott in the Hall-Scott Motor Car Company. The company manufactured locomotives, gas-driven railway coaches, interurban car bodies for electric railways, and automobile and airplane engines. By 1913, aircraft equipped with Hall-Scott four- and six-cylinder engines were flying in the U.S. In 1915, the Hall-Scott six-cylinder A-5 engine was being sold as a military engine in the U.S. and abroad.

Hall was commissioned a major in the Signal Corps in October 1917 and was promoted to lieutenant colonel in April 1918. In addition to co-designing the Liberty engine, Hall adapted the Le Rhone engine to U.S. production techniques, served as a troubleshooter in starting U.S. production of the DH-4 aircraft, and in October 1918 became chief of the Air Service's Technical Section in France. He received the Distinguished Service Medal for his wartime service, a sharp contrast to Vincent's WWI "reward" -- Hughes' recommendation that he be court-martialed. Dickey, 10-11, 96.

[65] Dickey, 12-13, 95. Sweetser, 175. McFarland, 1846-1849, 1891-1894.

[66] Dickey, 13-14. McFarland, 1895, 2041-2044, 2148-2150.

[67] Dickey, 14-15. Sweetser, 175-176. McFarland, 1895, 1941-1950, 2044-2050, 2150-2155. Foulois, 152-153.

[68]   Dickey, 15-16. McFarland, 2050-2052, 2155-2156.

[69]   Dickey, 16-17. McFarland, 2156-2157.

Loening, critical of everything the "Detroit gang" did, described the 8-cylinder Liberty as "a failure," "wrong," and "a complete washout." On the other hand, he conceded that the "redesigned Liberty" 12 was "the one that in time became a great engine all over the world." Loening, Our Wings, 81-82.

[70]   Dickey, 17. Sweetser, 176. McFarland, 1899, 1901, 2053, 2055-2061, 2158.

[71]   Dickey, 33-35. Sweetser, 177. McFarland, 2068-2079.

[72]   Dickey, 35-37. Sweetser, 180. McFarland, 1901, 1081-2101, 2158.

[73]   Emmons started his career as a lawyer in Detroit and became an outstanding industrial executive. He was detailed by the Navy to the Equipment Division for the engine production job. He influenced production as much as anyone and received the Distinguished Service Medal based on the recommendation of the War Department. Emmons built up an organization of 23 engine construction plants and 79 parts factories, and he was instrumental in getting an A-1 priority to cover the machinery, equipment, and facilities needed for engine production. Dickey, 59-60. McFarland, 1833-1835, 1901.

[74]   Dickey, 37-41, 95. McFarland, 1902, 2158-2159.

After testing of the number 5 and 12 engines, Hall was assigned to getting the American-built DH-4 into production and had no further contact with the Liberty engine.

[75]   Dickey, 42, 95. Sweetser, 177, 180. McFarland, 2101-2109, 2159.

[76]   Dickey, 61-62. McFarland, 884-885, 1906.

[77]   Dickey, 62, 65. Sweetser, 183-184. Foulois, 154-155.

[78]   Dickey, 66. Sweetser, 182.

[79]   Dickey, 19-23, 66-67, 92-99. Gorrell, 70-71. Sweetser, 179. McFarland, 1900, 1903, 1906-1941, 1950-2040. Goldberg, 18. Purtee, 107.

Dickey describes the various engines. Thus, the L-4, which was an experimental engine, weighed 398 pounds and developed 102 horsepower at 1,400 revolutions-per-minute, but would have developed more power with a regular carburetor. The L-4 was not produced in quantity because its power was suitable only for trainer aircraft and sufficient Curtiss, Hall-Scott, and Hispano-Suiza engines were available for that purpose.

The L-6 weighed between 540 and 560 pounds and developed between 200 and 215 horsepower which could have increased to 230 to 240 horsepower with refinements. However, the L-6 was not produced because it was too large for trainers and enough engines were already available for single-seat fighters. L-6s built by Thomas Morse and Wright, were considered for installation in Caproni triplanes, light bombers, blimps, and other aircraft.

Like the smaller versions, the L-8 had to compete with proven engines, in this case, the Hispano-Suiza-300. The L-8 also experienced severe vibration, forcing Buick to stop production after 15 engines. The L-8, the first of the Liberty series, was built in 21 days and was received at the Bureau of Standards on July 3, 1917. In July 1918, that engine was put on display by the bureau.

The U.S. agreed to supply the British with 11 percent of the Liberty engines produced. Thus, the British received 2,252 of the 20,478 produced, but only 980 were delivered prior to the armistice. Similarly, the French received 3,575 engines, but only 405 before the armistice. The settlement with Britain amounted to $16.6 million for engines and spares. France paid $21.3 million for engines and 3,310 sets of spares. Assuming Britain received spares equal to her quota of engines, the cost per engine and one set of spares was over $7,300. France paid less than $6,000 each for her engines, including spares.

The 20,478 engines were built by Packard 6,500; Lincoln 6,500; Ford 3,950; General Motors (Cadillac and Buick) 2,528; and Nordyke and Marmon 1,000. These production figures were against initial procurement of 6,000 from Packard; 6,000 Lincoln; 5,000 Ford; 3,000 Nordyke and Marmon; 2,000 General Motors; and 500 Trego Motors Corporation. One Ford innovation was cylinder forgings made from steel tubing, resulting in great saving in costs and quantity production. Ford also developed a special process for producing bronze-backed, babbitt-lined bearings in the crankcase and connecting rods. Ford's electric butt welding of the inlet and exhaust elbows to the top of the cylinder forging improved that process.

Mixter and Emmons (pp. 29-30) give the figure of 32,420 engines produced by November 29, 1918, dividing production as follows:

OX-5	8,458
Hispano-Suiza	4,100
Le Rhone	1,298
Lawrance	451
Gnome	280
A7A	2,250
Bugatti	11
Liberty	15,572

[80]   Dickey, 23.

[81]   Sweetser, 184-185.

[82]   Dickey, 23-24.

Inversion of a Liberty engine was first tried in December 1918 in an experimental 24-cylinder X-type model. This version was a combination of two L-12 models, the cylinders on one standing upright and the other pointing downward. The junction between the two engines was specially designed. The engine failed when a connecting rod broke. Further experimentation stopped, presumably because Vincent left government service on November 30, 1918. Dickey, 28.

The advantages of inversion caused continuing interest in converting the Liberty. Shifting the weight of the installed engine would make the thrust line coincide with the aircraft's center of gravity. Mechanics could work on the engine from the ground rather than on a maintenance stand. Relocation of the cylinders under the engine provided maximum visibility for the pilot and, when the inverted engine was air-cooled, the air scoops did not hinder visibility. Although inversion increased the weight of the engine, it also increased power. The next inversion test, made in February 1919, showed that the approach was feasible if oil scavenging and water flow problems could be solved. Dickey, 28-29.

In 1923, another inversion test occurred successfully but the oil had a tendency to overheat. The inverted engine. developed 422 horsepower at full throttle with a fuel consumption of .499 pounds per horsepower hour. At 90 percent of normal speed under a propeller load, oil consumption was 8.7 pounds per engine hour. The engine weighed 915 pounds with the generator but without the starter, air intake pipe, or exhaust pipe. Initial testing occurred in a DH-4B on September 5, 1923. Grover Loening was interested in the inverted Liberty engine, designing his amphibian for it. Twenty inverted Liberty engines were built for the Loening Amphibian, COA-1, in the Air Service inventory. The Allison Company of Indianapolis, Indiana, was the contractor for converting engines; the price per conversion was $1,472. Dickey, 29-30. Loening, 136-138.

[83]   Dickey, 24.

[84]   Dickey, 24-25.

[85]   Dickey, 25-28.

[86]   Dickey, 28. McFarland, 1896-1899.

[87]   Dickey, 30.

Vincent chose an epicyclic gear-reduction design because it used only known, successful construction. After studying the Rolls-Royce epicyclic gear, Vincent simplified its design and reduced the spur gear's weight 50 to 75 pounds. Initial tests indicated faulty lubrication and improperly set bearings. After fixing these problems, the assembly was tested successfully and Allison Experimental Engineering Company produced several sets. The Navy was especially interested in the geared version, but Vincent was concerned that emphasis on it would interfere with production of the direct drive engine. By February 1925, however, the Navy had a surplus of the geared engines and the Air Service in 1926 requisitioned them from the Navy for use in transports and bombers as air-cooled engines. Dickey, 30-31.

[88]   Dickey, 45-54. McFarland, 2160-2164.

[89]   Dickey, 52-53. Sweetser, 163-164. McFarland, 1547-1563, 1575-1596. Frey, 135-136.

Castor oil, particularly useful in rotary engines, did not congeal in cold temperatures at high altitude as readily as other oil and it was not seriously affected by gasoline entering the crankcase. Castor oil did release fumes in flight. Pilots inhaling these fumes for any length of time experienced the same physiological effect as an oral dose of several spoons full of castor oil. Reportedly, pilots either developed immunity over time or drank quantities of blackberry brandy to offset the effects. Mineral oil was not much better. Frey, 135.

[90]   As quoted in Dickey, 53-54, and McFarland, 1555. See also, Sweetser, 164-166. Purtee, 47. Frey, 135.

Many of the "pet" oils used by pilots had an operational life of only five to ten hours. A system for reclaiming Liberty Aero Oil was developed during the war, enabling the oil to be reused several times. More than 162,000 gallons were reclaimed. Frey, 135-136.

[91]   Dickey, 68-69. Walker-Wickam, 183. Arnold, 98. Foulois, 152-153.

92   Dickey, 69-70. Bilstein, 36-37. Claussen, 18. Goldberg, 16. Holley, 59-60.

Clark argued that systematic bombardment could end the war sooner than sending another 1 or 2 million men to the trenches. He called for bombing the enemy into exhaustion. Similarly, Major Edgar S. Gorrell advocated systematic bombing of Germany to destroy moral and materiel. Holley, 135-136, 165-169.

Most American aerial units overseas flew into combat in French Spads and Nieuports. Even with borrowed planes, the brief American combat record in the air was impressive. The first aerial victories came on April 14, 1918, but American squadrons shot down an estimated 850 planes and balloons by the armistice in November. Bilstein, 37.

93   Dickey, 70-71, 102. Gorrell, 34, 65-68. Sweetser, 190-199, 246-248. Goldberg, 17. McFarland, 1708-1726. Loening, Our Wings, 98.

Holley's thesis that doctrine -- or its absence -- influenced the development of weaponry was reflected in his critique of the U.S. decisions to build the Liberty engine and the DH-4 aircraft. He regarded the Liberty engine as a microcosm of the U.S. wartime struggle to develop aircraft: standardization was in conflict with development of superior weaponry. Only with the Liberty engine did the U.S. succeed in balancing mass production and improved performance. Holley, 118-131, 151-152, 175-176:

> *The decision to develop a new and superior engine rather than copy Allied designs was of itself a decision favoring quality over quantity. It is true that the Liberty engine was subsequently blamed for many of the failures of the program for aviation when it proved difficult to convert airframes to mount the new standardized engine. . . . Unfortunately for the success of aircraft production in the United States, the decision to build superior engines was apparently not made as a logical consequence of a conscious and widely accepted policy stressing quality over quantity. . . . That it did not result from conscious policy is suggested by the parallel decision to put the inferior and obsolescent DH-4 rather than the DH-9 into mass production. Here, side by side, were two violently conflicting policies. One favored better weapons, the other favored more weapons.*

Holley argued that the U.S. entered the war without a clearly defined doctrine of aerial warfare and that this limited doctrine unduly favored observation and ground support over strategic air power. Thus, that policy resulted in production of observation aircraft rather than bombers during the war, and then the wartime experience was used by the General Staff as "proof" that aviation should be ancillary to Army operations -- a self-perpetuating argument that hurt the post-war Air Service. In fact, after the war the General Staff concluded that field artillery was more effective and economical than aircraft, arguing that it took two squadrons of bombers to equal the work of one 155-millimeter gun! Holley, 157-168, 171-172, 176-177.

Yet by the summer of 1918, over 8,000 military aircraft were on the western front. When the allies moved against the Germans at St. Mihiel in September 1918, Colonel William Mitchell directed the largest air force yet organized -- almost 1,500 fighter and bomber aircraft -- in a coordinated offensive. Mitchell deployed 500 fighters and bombers over the front lines while two waves of 500 aircraft attacked the German rear, destroying supplies, communications centers, and transportation. Bilstein, 34. Arnold, 81, 158. Foulois, 159-160, 176-182.

94   Dickey, 71, 103. Gorrell, 34. Sweetser, 203-209, 242-244.

[95] Dickey, 73-74.

The first aerial crossing of the Atlantic was made by Navy Commander Read in May 1919 in an NC4 flying boat powered by four Liberty engines. In 1924, Douglas World Cruisers powered by L-12s completed the first aerial circumnavigation of the world, a flight of 26,345 miles in 363 hours of flying. Dickey, 80.

[96] Dickey, 32.

Loening used inverted Liberty engines on his amphibians. Loening, Our Wings, 124, 127, 131.

[97] Dickey, 74-78, 106. Holley, 121, attributed the ban on using Liberty engines to congressional directive.

The Army's experience with the Curtiss D-12 indicated that the cost of overhauling it and the Liberty were comparable. The higher initial cost of the D-12 was offset almost entirely by its longer life and cheaper operation. Purchased and overhauled in large quantities, the cost of the D-12 should have been less than the Liberty. The number of maintenance hours expended on the D-12 was about half of those expended on the Liberty, mainly because of accessibility. The D-12 was also regarded as more reliable since fewer crashes occurred because of engine failure.

[98] Dickey, 78-80. Foulois, 203-204.

[99] Purtee, 75-76, 83. Mooney & Layman, 30. McFarland, 633, 699-701, 2065. Frey, 117-118. Sweetser, 212-218.

[100] Purtee, 78-79, 83, 103. Dickey, 68. Sweetser, 245. Arnold, 67-71, 113.

Loening further argued that at the end of the war, largely because of the policies of Ryan and Potter who succeeded the automobile group, the U.S. had four outstanding aircraft ready for production: the Martin bomber with Liberty engine; the Thomas-Morse pursuit; the Loening two-seat fighter; and the Vought VE-7. For the specifics of Loening's thesis against the "Detroit gang," see his two books: Our Wings Grow Faster (Garden City, NY: Doubleday, Doran & Co., Inc., 1935), 69-72, 74-77, 81-92, 95-96, 99-100; and Takeoff Into Greatness: How American Aviation Grew So Big So Fast (New York: G.P. Putnam's Sons, 1968), passim. Loening went abroad in April 1917, three days after the U.S. declared war, and returned at the end of June -- in time to warn the Aircraft Board that its optimistic predictions were unlikely to be realized. While abroad, Loening learned much about British aviation progress.

[101] Purtee, 79-80.

[102] Purtee, 81-82, 84, and Appendix D. Mooney & Layman, 31, 34. Sweetser, 220-221. McFarland, 635, 701-713. Frey, 112-113. Sweetser, 211-212, 218, 220-221.

[103] Biographical files in the Air Force Museum show that Colonel Thurman Harrison Bane (June 12, 1884-Feb. 22, 1932) was another accomplished individual. Born in San Jose, California, Bane was graduated from West Point in 1907, and transferred to the Signal Corps in November 1916. He served with the Aviation School at North Island from 1916-1917. While at the school, he drew up a course in aeronautics and design based on articles then being published in Aviation magazine by Alexander Klemin and Thomas H. Huff, and was placed in charge of the experimental shops there. Late in 1917, he transferred to Washington, D.C., serving on the Joint Army and Navy Technical Aircraft Board and as Executive Officer with the Air Division of the Signal Corps.

In May 1918, Bane became chief of the Technical Section of the Division of Military Aeronautics when it was established. His duties required him to coordinate on aircraft specifications with the Bureau of Aircraft Production's engineers, and he transferred to Dayton in late 1918 to act as liaison. After the armistice in November 1918, he became responsible for experimental engineering at McCook Field and commanding officer. He was in charge when the Engineering Division formed and led it through the postwar transition to stability and success. Among his achievements was the establishment of the Air Service Engineering School at McCook, which opened in November 1919. He served as the first commandant of the school. He retired because of ill health on December 15, 1922. At the time of his death, he was vice president of American Airways, Incorporated. He died of a brain tumor.

[104]   Purtee, 82, 95-96, 106. Mooney & Layman, 31-33. Sweetser, 217-219, 237-238. McFarland, 30, 33-34, 97-102, 288-289. Frey, 119. Goldberg, 15. Holley, 68-69. Foulois, 157-158.

[105]   The following discussion is based on McFarland, 403-407, 411-419, 580, 595; and Claussen, 19-20.

Before U.S. entry into World War I, the National Academy of Science formed a National Research Council to promote general scientific research. During the war, the council became directly involved in military issues, especially aircraft and signaling problems. Because the NRC's funding and personnel were inadequate for such research, the Chief Signal Officer in July 1917 established a Science and Research Division in the Signal Corps, using the NRC's Physics Committee as advisory agent. The division was placed under two civilian scientists in uniform: Lt. Col. Robert A. Millikan, a distinguished physicist, and Dr. (Major) C.E. Mendenhall. The division was formally constituted on October 22, 1917. At first, the division was given charge of all R&D work assigned by the CSO, a vague assignment. But the idea was that, as problems arose, the division would detail men to work under the direction of the military divisions most concerned.

In fact, the Science and Research Division was not fully constituted until February 1918. At that time, the CSO gave the division greater independent responsibilities, charging it with investigation of all developments in science and invention, with furnishing the other divisions information important to their duties, with performing all scientific research for the Signal Corps, and with training of all meteorological personnel. This constituted the first attempt to organize research for military purposes in the U.S. The armistice came nine months later; thus, only a few of the division's results had a part in the actual military operations abroad.

In May 1918, the Science and Research Division transferred to the Bureau of Aircraft Production. The division's duties under the BAP included scientific investigation and research work for both the Division of Military Aeronautics and the Signal Corps. At the time of the armistice, the personnel included 22 commissioned officers, 120 enlisted, and 16 civilian scientists of the Air Service, and about 15 officers and 150 enlisted men of the Signal Corps principally engaged in meteorological problems and training of meteorologists for foreign service. The work of the department was carried out in five sections: signalling; aeronautic instruments; bombsights; chemical; and meteorological. A photographic section in the division became a separate department of the BAP in July 1918. An aerodynamical research section also transferred to BAP's Airplane Engineering Division.

The Science and Research Division offices were located in Washington, D.C., but the actual work occurred at many places, according to the physical facilities offered. The list of government departments, educational institutions, and commercial laboratories that cooperated from time to time was a long one. The division built a small experimental laboratory and instrument shop at Langley Field for work in photography and instrument testing. A small station established at Johns Hopkins University enabled balloons and signalling testing. Much shop work was done at Carnegie Institute in Pittsburgh, and most of the experimental work that did not require flying was done at the Bureau of Standards or the Bureau of Chemistry.

The Aeronautic Instruments Section of the Science and Research Division was charged with the writing of initial specifications on all aeronautic instruments. Among the more important ones dealt with by the section were a compass in use by all American aircraft; the air speed indicator on American aircraft; a new and improved venturi-pitot tube for use with air speed indicators; and an altimeter. Other work was done on a gasoline level gauge. Dr. G.S. Fulcher invented a leak-proof gasoline tank for fighting aircraft to reduce the danger of fire from incendiary bullets. The tank was developed in collaboration with the Miller Rubber Company and the S.F. Bowser Company. An aircraft carburetor, designed by Mr. L.S. Marks, was completed at the time of the armistice, but had not yet been tested. The carburetor was designed to relieve the pilot of all control of the engine except through the throttle. Very valuable studies of aircraft fuels were made at the Mellon Institute in Pittsburgh by Dr. Rosanoff. The fuel testing showed considerable increase in efficiency at high altitudes. Other problems addressed included double propeller engine of special design. Wind tunnels were built by the aerodynamical research section, one at Bridgeport, Connecticut, and one at the Bureau of Standards. The wind tunnel work was later transferred to the Airplane Engineering Division.

After the armistice, most of the staff returned to civilian life and uncompleted projects were taken over by the Engineering Division of the Air Service. Holley charged that this restructuring subordinated science to engineering in the Air Service, thereby harming fundamental research. By emphasizing production over experimental development (at the armistice 75 percent of the people in the Engineering Division were concerned with production problems) science was neglected -- even scientific liaison was interrupted. Holley, 112-117, 153-156.

[106] Blee, 15-19. Purtee, 95-96. McFarland, 37, 288-292. Frey, 120-121.

[107] Blee, 19-26. Purtee, 96-97. Mooney & Layman, 35. McFarland, 35-36, 292-298. Frey, 121-122.

[108] Blee, 26-27, 33. McFarland, 298-299. Frey, 123.

[109] Purtee, 97, 106. McFarland, 37-38. Frey, 122-123. Sweetser, 238. Goldberg, 15. Holley, 68-69.

[110] Blee, 27-28, 41. Purtee, 102, 106, and Appendices B and C. McFarland, 299-300. Frey, 120, 123.

[111] Blee, 28-30, 33-36. Purtee, 102, and Appendix B. McFarland, 300-301, 304-305. Frey, 123-124.

[112] Blee, 38. McFarland, 309. Frey, 124.

Purtee, page 102, and Claussen, page 16, give different figures for the total personnel at McCook at the armistice in November 1918 as 2,240: experimental engineering personnel at McCook numbered 14 officers and 1,335 civilians; production engineering had 26 officers and 330 civilians; and business and military had 18 officers, 250 civilians, and 267 enlisted men.

[113] Mixter and Emmons, 4-5, 10. Bilstein, 37. Goldberg, 18. Holley, 106. Purtee, 104. Claussen, 21.

[114] Blee, 41. Purtee, 108, and Appendix C. Mooney & Layman, 36. McFarland, 273, 312. Frey, 126.

[115] Purtee, 108. Mooney & Layman, 36-37. McFarland, 635.

[116] Purtee, 109-110, 122. Arnold, 91.

Brigadier General Mitchell retained his rank and became Assistant Chief of the Air Service, a development that outraged Foulois who again became a major. Foulois, 185-186.

[117] Purtee, 112.

[118] Purtee, 113-114, 123. Frey, 188-189. Walker-Wickam, 99. Aircraft Year Book 1920, 282.

[119] Purtee, 114, Frey, 189.

[120] Purtee, 114-115, 123. Frey, 189.

[121] Purtee, 152-153. Patrick, 82-85. Walker-Wickam, 100. Aircraft Year Book 1922, 167-169; 1923, 278-279; 1924, 294-295; 1925, 259-261; 1926, 273-275; 1927, 306-307.

General Patrick established eight sections in the Engineering Division on December 1, 1921, including planning, technical, factory, flying, procurement, supply, patents, and military. However, by 1923, charts show many more sections developed. By the mid-1920s, the Engineering Division called all of its subdivisions "sections." Thus, a planning section reported to the division chief; seven sections were under the chief engineer: airplane, armament, equipment, lighter-than-air, material, power plant, and service liaison; one section was under the chief production engineer: production; and -- eventually -- more than a dozen sections were under the assistant chief of the division, including comptroller, finance, factory, flying, employment, legal, patent, supply, technical data, inspection, maintenance, and quartermaster. Invariably, each section was made up of several branches; the Power Plant Section, for example, had 10 branches by 1925. Establishment of the Materiel Division on October 15, 1926 within the new Army Air Corps changed this alignment, but not the work of the former Engineering Division which was reduced to an Experimental Engineering Section.

[122] Blee, 84-85. Purtee, 102. McFarland, 360-361.

[123] Blee, 85. Purtee, 102. McFarland, 361.

[124] Blee, 85-86. McFarland, 362.

[125] Blee, 86-87. McFarland, 367.

[126] Blee, 87-88. McFarland, 367, 371.

[127] Blee, 88-89. McFarland, 372, 374.

[128] Blee, 89-90. McFarland, 374, 376.

[129] Blee, 90. McFarland, 376.

[130] Blee, 91. McFarland, 379.

[131] Blee, 91-92. McFarland, 379, 382.

[132] Blee, 92. McFarland, 382.

[133] Claussen, 18-21.

[134] McFarland, 1706. Blee, 41-42. McFarland, 313. Frey, 126. Walker-Wickam, 183-184.

135 McFarland, 1702, 1704, 1726-1733. Blee, 42-43. McFarland, 313-314. Frey, 127. Walker-Wickam, 184-185.

136 McFarland, 1701, 1704-1705. Blee, 43-44. McFarland, 314-315. Frey, 127. Walker-Wickam, 185.

137 Walker-Wickam, 185-186. McFarland, 1703.

138 Blee, 61-63. McFarland, 335, 340-341. Frey, 129. Hallion, 51-52. Loening, Our Wings, 85-86.

139 McFarland, 1805-1811. Walker-Wickam, 186-187. Frey, 129.

140 McFarland, 1703, 1801-1804. Blee, 44. McFarland, 315. Frey, 127, 129. Walker-Wickam, 185-186.

141 McFarland, 1703, 1811-1814. Mixter & Emmons, 51. Flight, 91. Loening, 85-92, 102, 104-107.

142 Blee, 44-45. McFarland, 315-316. Frey, 127.

143 Blee, 45-46. McFarland, 316-317. Frey, 127.

144 Blee, 53-54. McFarland, 328-329.

145 Blee, 51-53. McFarland, 326-327. Frey, 127-128.

146 Blee, 54-56. McFarland, 329-330.

147 Blee, 56-57. McFarland, 330-332.

148 Blee, 53. McFarland, 327-328. Frey, 128.

149 Blee, 46-51. McFarland, 317-326. Frey, 127.

150 Blee, 58-61. McFarland, 332-335. Frey, 128-129.

151 Blee, 63-68. McFarland, 342-345. Frey, 130.

152 Blee, 68-71, 75-78. McFarland, 346-349, 352-355. Frey, 130-134.

153 Blee, 71-75. McFarland, 349-352. Frey, 131.

CHAPTER III

McCOOK FIELD, MARCH 1919 - OCTOBER 1927
POWER PLANT SECTION

The Great War left the Air Service with an experimental engineering laboratory for aviation at McCook Field. After the armistice, however, the laboratory's peacetime role was unclear. Also, McCook Field's temporary nature remained unresolved. Under the leadership of Colonel Thurman H. Bane and his successors, the Engineering Division that emerged in March 1919 fostered a scientific rather than a military environment. The division made incredible progress before McCook Field closed in October 1927. Indeed, for the division's Power Plant Section the 1920s proved to be a kind of "golden age" despite limited resources.

Severe reductions in funding and staffing followed soon after the war. During the ensuing decade, many of the division's key personnel went to private industry, taking with them the knowledge and techniques of McCook Field. The division, responsible to Air Service headquarters in Washington, was almost autonomous because of its expertise. No other Army organization had the aviation capabilities of McCook Field. The people at the Engineering Division learned every day from their mistakes and their successes. In the process, they established a business-like organization that encompassed the whole of aviation: engineering, material analysis, procurement, quality control, test procedures, flight testing, and training.

McCook Field usually had on hand between 70 and 90 active aircraft representing 50 types. In addition to aircraft, engines, and equipment, the division contributed procedures, tests, standards, and materials, including manufacturing processes. Other contributions ranged from basic instructions to complex formulas and specifications. The division set precedents for writing contracts, fostering competition, preventing fraud, and ensuring performance.[1]

The Air Service to 1926

Major General Charles T. Menoher, appointed chief of the Air Service in December 1918, succeeded John D. Ryan. He continued in that position until October 1921. General Menoher was an infantryman, not a pilot. Although he regarded aircraft as the most potent weapon ever produced, he opposed establishment of a separate air force.[2] Brigadier General William (Billy) Mitchell, who had succeeded General Kenly as head of the Division of Military Aeronautics in March 1919, subsequently became assistant chief of the Air Service. The opposing views of Generals Menoher and Mitchell inevitably caused conflict between them.[3]

To cover fiscal year 1920, congress appropriated approximately $25 million of the $55 million the Air Service requested. For the next business year, General Menoher requested $60 million but the amount only increased to $33.435 million. In 1922, funding reduced to $19.2 million. In 1923, the appropriation cut to $12.895 million and continued near that level for the next four years.[4]

Based on the funds the Air Service received to operate, the General Staff authorized a severely reduced Air Service of 1,000 officers and 11,000 enlisted by the end of 1920. The action removed at least some of the uncertainty surrounding the status of the postwar Air Service. However, with the Army's appropriations for business year 1920 cut by congress, Secretary of War Newton D. Baker ordered all temporary aviation officers discharged by September 30, 1919. The discharged officers could remain as civilian employees.[5]

Efforts by airmen to achieve a separate department of aeronautics in the aftermath of WWI did not succeed. Instead, congress passed the Army Reorganization Act of June 1920. This legislation permitted the Army to keep its air arm as an integral part of its ground organization rather than having it revert to the Signal Corps. As part of the Army's combat line, the Air Service had a measure of autonomy comparable to the artillery, cavalry, or infantry. The Air Service organization included two wings with headquarters at Kelly and Langley Fields. The act of 1920 increased the authorized number of military in the Air Service to 1,516 officers and 16,000 enlisted of the Army's total authorization of 280,000. The actual strength, however, in September was only 1,170 officers and 7,846 enlisted. The Air Service in June 1923 had 867 officers, an authorized enlisted personnel of 8,764, and 3,484 civilian employees. By 1926, with Army strength at 134,938, the Air Service included only 919 officers and 8,725 enlisted.[6]

The reorganization of 1920 remained in effect during most of General Menoher's tenure. General Menoher resigned as chief of the Air Service in protest against General Mitchell using the prestige he won in World War I to agitate for public support of an independent air force. On October 5, 1921, Major General Mason W. Patrick succeeded General Menoher as chief. He held the position until his retirement in December 1927. During his tenure, Patrick qualified as a pilot in June 1923 at age 60. General Patrick's position on an independent air force supported Mitchell's, but the new chief was far less confrontational and much more pragmatic in his approach.[7] In December 1921, General Patrick restructured the Air Service into five divisions: Personnel, Information, Training and War Plans, Supply, and Engineering. In this restructuring, the Engineering Division's organization aligned into various sections.[8]

Engineering Division, 1919 - 1926

The Engineering Division emerged from WWI with the mission[9]

> to design, develop, and test all airplanes, airplane engines, accessories, and materials to meet the requirements of the Air Service; to prepare production drawings, specifications, and, where necessary, models of all aeronautical equipment for production; and to assist and supervise the experimental and production manufacture of all aeronautical equipment being designed and constructed for the Air Service by the aeronautical industry.

Thus, during the 1920s, McCook Field's technical sections focused on designing, developing, and testing aircraft, engines, and related equipment. The Flying Section conducted thousands of test flights in 1919 alone.[10] The Airplane Section worked on 16 types of aircraft (later consolidated to five in 1925 to pursuit, bombardment, observation, training, and cargo aircraft.) Both liquid- and air-cooled engines and variable- and reversible-pitch propellers powered the aircraft. The Armament Section adapted various types of machine guns, flexible mounts, armament installations, synchronizing devices, aircraft cannon, bombs and bombing equipment to aircraft use. The Equipment Section concentrated on a variety of instruments, clothing, and installations. They included parachutes, leak-proof and auxiliary gas tanks, flotation gear, modification of DH-4 aircraft for photography, a gyro compass, a portable field engine cranker, a pressure fire extinguisher, and a central electric power plant. The Material Section tested many different materials for aircraft, engines, and equipment. The Balloon and Airship Section, added in 1921, dealt with developments for lighter-than-air equipment. The Power Plant Section worked on engines for aircraft and airships, including refining the Liberty and Hispano-Suiza engines, and developed new liquid-cooled and air-cooled engines, superchargers, test equipment, cooling systems, and fuel systems. Until McCook Field closed in 1927, these sections were responsible for the engineering work.[11]

Despite its broad mission, the Engineering Division continually experienced reductions in personnel and funding reduced steadily during its existence, seriously undercutting the development of aircraft, equipment, and accessories.

Staffing Reductions. Just prior to the signing of the armistice, the Airplane Engineering Division included 58 officers, 385 enlisted, and 1,915 civilian personnel, By January 1, 1919, McCook Field personnel declined to 1,474 people. By January 1, 1920, the Engineering Division's work force fell to just 286 military and 1,061 civilians.[12]

McCook Field lost so many people that concern about its ability to sustain aviation progress arose. In 1920, concern for national security in the future caused one observer to call for stronger support of aviation in terms made familiar by repetition ever since:[13]

The development of aeronautical engineering as carried on by institutions like McCook Field is not a cheap proposition. On the other hand, however, it does not require any sums that might in the least prove to be burdensome to a country as wealthy as ours, and the value of such an investment would be repaid a thousandfold in many ways, especially if we should have the misfortune, together with the rest of the world, to be faced with a grave war emergency in the future.

McCook Field's civilian employees numbered 1,573 in July 1921, the peak of its staffing in the postwar period. The number fell off considerably thereafter along with appropriations. Nonetheless, the Engineering Division continued to attract hordes of aspiring employees. More than 11,500 applicants applied between January and September 1921. In fact, the division was one of the largest employers in Dayton. As Slipstream, its quasi-official periodical, pointed out: *"McCook Field can point with pride to her stability and ranks third in the city as to numbers in personnel and salary payroll, being outclassed only by the NCR and Delco."*[14]

Personnel cuts proved difficult to absorb. To reduce expenditures, the Engineering Division cut payroll by $40,000 between August 26 and September 30, 1921. The division initiated a survey contemplating a general adjustment of employee salaries to come within the allotted payroll that ultimately resulted in 200 employees losing their jobs. Explained Major Lawrence W. McIntosh, acting chief of the division at the time:[15]

This reduction has been accomplished by the suspension on leave without pay, of sufficient employees, for an indefinite period, with definite information to such employees within sixty days as to whether their suspension is to be a termination of their services by reduction of force.

Commenting on work done in 1922, the assistant chief lamented in words numerously repeated in one form or another through the ensuing years in Dayton:[16]

This work has been accomplished in spite of physical obstacles and a reduction in the Engineering Division appropriation which necessitated the discharge of 136 employees. This unfortunate reduction of personnel at the middle of the year necessitated a reorganization of groups engaged on various projects and unavoidably interfered to a considerable extent with the progress attained.

In July 1921, the Engineering Division was down to 109 officers and enlisted men and 1,541 civilians; by 1922, the division's civilian personnel declined to approximately 1,300. During fiscal 1923, the number of civilians in the Engineering Division decreased to 1,180, a number half McCook Field's personnel strength at the end of WWI. That number tailed off in business years 1924 and 1925. The money available to the Engineering Division also fell off through its existence.[17]

Not surprisingly, many employees accepted offers from private industry. Following a spate of resignations in 1923, the Engineering Division's quasi-official periodical commented:[18]

> *It is interesting to note that in most instances of resignations listed here the reason given for the move is "to accept a better position." This points out that the lack of sufficient appropriations provided by Congress to carry on the Engineering work of the Air Service makes it extremely difficult for the Government plant to cope with conditions which have so greatly improved in all lines of commercial industry, and which make promising offers for specialized services such as are required at McCook Field.*

On the other hand, McCook Field's employment was actually large, given the number working in the civilian aircraft industry. During the congressional debate over appropriations for aviation at the beginning of 1925, employment in the U.S. civilian aircraft industry was not more than 1,500 persons.[19]

Funding Reductions. The reductions in personnel followed sharply reduced appropriations. For fiscal year 1920, the Engineering Division at McCook operated with less than $6 million of the Air Service's $25 million. Thereafter, appropriations steadily decreased, reaching $5 million of $33.435 million in 1921; $4.3 million of $19.2 million in 1922; $3.5 million of $12.895 million in 1923; and $3 million of $12.426 million in 1924. After that, Air Service appropriations began to rise, but the R&D budget continued to decline, reaching its lowest point in fiscal 1927.[20]

Fiscal Year	$ (Millions)
1920	$6.0
1921	$5.0
1922	$4.3
1923	$3.5
1924	$3.0
1925	$3.0
1926	$2.7

Adding procurement to these figures, however, increased the total considerably. The Air Service provided $53.21 million for research, development, and procurement from 1921 through 1926, 40 percent of its total obligations for that period.

The experimental appropriation for fiscal year 1924 was $3 million of which the Engineering Division received $2,819,224.15 for experimental purposes. Of that amount, $2,787,364.59 went to payroll, operating expenses, and material, and for purchasing experimental aircraft, engines, and equipment. Another $29,863.33 transferred to the Bureau of Standards, Forest Products Laboratory, and for purchasing equipment from foreign nations. The balance of $1,996.23 remained unspent at year's end.[21]

The appropriation for experimental and research engineering for fiscal year 1926 was $2,730,000, a decrease of $350,000 compared to the previous business year. Of the total, $2,644,900 went to the Engineering Division with a limit of $1,730,000 for civilian pay.[22]

Yet, when the Lampert committee of congress reported its findings in December 1925, the committee calculated that from 1920 to 1924, U.S. expenditures for aeronautical purposes had amounted to $424,234,107.90. The committee also lashed out at the military's expenditures for aviation R&D:[23]

That the expenditure of more than $30,000,000 in the five-year period by Army and Navy combined, for so-called experimental and research work, has built up or maintained in both bureaus a governmental aviation industry larger than the entire civilian industry, employing at the naval aircraft factory at Philadelphia at the time of the investigation approximately 1,100 men, and at McCook Field and Fairfield approximately 1,200 men, and entailing a naval aircraft factory pay roll of approximately $2,200,000 a year, and at McCook Field approximately $2,000,000 a year; that these expenditures for experimental and research work have not produced results commensurate with the expenditures and that both services, since the committee began its hearings, have greatly curtailed these expenditures and have shown a strong tendency to recognize the fact that it is unwise for the Government to spend such a large proportion of the resources in this branch.

McCook Field's assessment of the adequacy of the government's funding for R&D differed greatly. The chief of the Engineering Division, Major Lawrence W. McIntosh, in his annual report for 1923 declared that the appropriations hurt basic research and aircraft development. He explained:[24]

The continued decrease of appropriations permits only the solving of the more important problems at hand in the quickest manner possible, but does not permit much endeavor along lines of fundamental research. The effects of the little fundamental research are already too apparent, as is indicated by failures due to unknown stresses of structure which are apparently of unsufficient strength to withstand the strain, due to ever increasing high speed of the airplane.

General Patrick, chief of the Air Service, echoed this assessment in his report to the Secretary of War:[25]

The continued decrease in appropriations permits the undertaking of only the most immediate and important problems and practically precludes the possibility of research work. The effects of this have already been made too apparent through failures due to unknown strains and stresses in high-speed airplanes. Fundamental research though a tedious process, and one which is fruitful only after long periods of patient application, is nevertheless the vital need and the surest means of real advancement in the science of aviation.

General Patrick described the critical conditions caused by restricted resources in these words:

The great majority of the aircraft now on hand were produced during the war, are rapidly deteriorating, and even when completely reconditioned have but a very short life. Furthermore, 80 per cent of the airplanes are of an obsolescent training type, unsuitable for combat purposes. It is absolutely essential that the purchase of new aircraft to replace that produced during the war and to offset the constantly increasing shortage be undertaken immediately. Since it requires about 18 months to secure delivery after a contract has been executed, it is apparent that no relief from the present situation can be expected before 1926. Appropriations now being made for the purchase of new aircraft are insufficient to meet the requirements of even the present inadequate peace-time establishment of the Air Service.

The inventory of active aircraft steadily decreased. Much of the postwar inventory consisted of equipment left over from WWI. At war's end, the Air Service had more than 3,500 DH-4 aircraft and almost 12,000 Liberty engines on hand. At first, nearly all replacement aircraft and equipment came from these stocks, and the Engineering Division modified existing models instead of developing new aircraft and engines. Wartime aircraft proved dangerous. During 1920 alone, 150 crashed.[26] People predicted that the number of available aircraft would decline from 1,970 to 289 by 1926. The number of these aircraft expected to be of WWI vintage was large: 1,531 in 1923 and 102 in 1926.[27]

In his annual report for fiscal year 1924, General Patrick lamented the unhealthy state of the aeronautical industry in the United States, especially the lack of a war reserve, the difficulty of quantity production in the event of war, and the weakness of commercial aviation. In describing this situation, General Patrick added:[28]

The war-produced stock has now been exhausted. Many planes have been rebuilt more than once. A careful technical survey which has been going on for eight months . . . indicates that the deterioration in some classes of equipment in storage awaiting rebuilding, notably foreign built and training planes, is greater than was expected. In the interests of safety, whenever the damage justified it, these planes have been dismantled, the unserviceable parts destroyed, and those parts which can be safely and economically used again, turned into stock for repairing other planes.

Despite the funding and staffing handicaps, however, the Engineering Division at McCook Field made considerable progress in improving the airplanes, engines, and accessories left over from wartime production, and in developing new equipment. In judging the practical value of its achievements, the Engineering Division measured itself against the standard used by General Patrick to have the most advances types of aircraft, engines, armament and other miscellaneous equipment ready for immediate production and service. The Engineering Division interpreted the needs of the Air Service and translated its requirements into specifications and other engineering data to use in procuring aircraft and engines for experimental purposes. The division needed engineering data such as drawings, specifications, parts lists, and physical model in readiness to facilitate quantity production of advanced aircraft, power plants, and equipment. McCook Field's leaders boasted that their approach advanced the progress of general aviation.[29]

In assessing the effect of its achievements on industry, the Engineering Division cited its role as the Air Service's central engineering organization. The division made the results of its tests and experiments part of the permanent record, publishing reports that recorded every step in the process. Most of these reports were readily available to industry, thus affording industry with the benefits of the division's research and development efforts. Unlike the Engineering Division, commercial organizations did not share their work with the industry. A company investing money and time on a development felt justified in keeping this information for itself. Examples of commercial developments thus reserved were duralumin tubing developed by Aluminum Company of America; and electric welding by the Boeing Airplane Company.[30]

The division's struggle to overcome its handicaps was successful to a remarkable degree. Despite limited resources, General Patrick in his annual report for 1923 expressed his appreciation for McCook Field's capabilities:[31]

> *The Air Service Engineering Division is probably the most complete, the most efficient, and the most productive aeronautical laboratory in the world. Its doctrine of progressive conservatism based on thorough research and rigid adherence to sound engineering principles has served to give it a credible share in practically every record of note now listed in the annals of aviation.*

Power Plant Section

The role of the Power Plant Section in the Engineering Division was critical to the growing success of aviation during the 1920s. Pointing to the importance of aviation for the future, Major Bane referred in 1922 to the role of engine development as a condition of success:[32]

> *I would rather accept a salary of $5,000 in the airplane industry than work for $50,000 in the automobile world. Not that I think automobiles are a back number, or soon to be, for I believe they will continue to be useful, but I do think they have reached the peak of their popularity. The coming thing is the airplane. I am just as sure of it as I ever was on any point. Of course, it may be some years before airplanes are accepted by the general public as a practical vehicle of conveyance, but they will be as soon as we get a reliable motor, and a lower landing speed. I do not think this will take long for much progress has been made on these improvements in the last year or two.*

Looking back at the development of the aircraft piston engine, S.D. (Samuel Dalziel) Heron concluded that the process was almost entirely empirical with mathematical analyses used only on a minor scale compared to development of the gas turbine. For example, Heron noted that Pratt & Whitney

Aircraft ended up redesigning its entire engine after the company succeeded in breaking every single item in the P&W R-2800. Thus,[33]

> *The progress of the piston aircraft engine may be justly described as a triumph of development. Most of the useful development has been the product of test bed running, and the process of break, burn, melt. The development process has almost always involved steadily increased power from a given engine type.*

This cut-and-try approach, observed Heron, was also fundamental in the development of aircraft fuels:[34]

> *Although a great deal of very able science has been involved in the synthesis and manufacture of piston engine fuels, the process has been empirical, since it was largely a question of trying fuels in engines. After engine test, we generally knew how fuels behaved but not why.*

The Organization in 1920. Major George E.A. Hallet organized the postwar Power Plant Section at McCook Field. He combined the propulsion activities at McCook Field with the testing elements from Wilbur Wright Field where he had served during the war.[35] By 1920, the Power Plant Section had four distinct branches: Engine Design, Dynamometer, Installations (including Fuel Systems), and Cooling Systems. A photograph published in the May 1920 issue of Slipstream gave some idea of the number of people assigned to the Power Plant Section. The photograph showed 64 people, including at least five of the secretarial staff. Only a few in the picture were in military uniform and a few men looked like they might have worked in McCook Field's shops. The Power Plant Laboratory, the test element within the Power Plant Section, included some 60 engineers, mechanics, and clerks under Charles Fayette Taylor and his assistant, Iskander Hourwich.[36]

The mission of the Engine Design Branch stated to design aircraft engines and their components. For this purpose, the branch supervised a designing room unit and the manufacture of parts. Designing engines necessitated extensive research and analysis of test data to solve design problems, careful study of existing engine designs, and calculations and stress analyses of new designs. Glenn D. Angle, who came to McCook Field in 1918 from the Curtiss Aeroplane and Motor Company, headed the branch until 1924. Theos E. Tillinghast assisted him and was responsible for research and calculations. W.O. Warner supervised design.[37]

The Engine Design Branch, which began operations early in 1919, achieved significant progress. The Engine Design Sub Unit in the Shop Engineering Section constructed a device dubbed a "rubber engine" because of its extreme adaptability to testing cylinders of various stroke-bore and value gear combinations. The single-cylinder test engine's crankcase design included an adjustable crankshaft and connecting rods of varying lengths. The engine made it possible to test cylinders having bore diameters from 4 to 8 inches and strokes from 4 to 10 inches. Compression ratios varied from 4-to-1 to 10-to-1.

Thus, the test engine covered the range of most existing aircraft engines, and enabled determination of the value of any cylinder design before constructing a complete engine.[38]

By 1920, the Engine Design Branch designed and built two universal test engines for testing and developing individual cylinders. The cylinders were both water- and air-cooled, and came in a range of sizes. It also designed and built adaptations for the universal test engine including the ABC (All British Engine Company, Limited) Dragonfly, the Liberty, and two sizes of the J.W. Smith radial engine cylinders. The branch designed and constructed a 5x6-inch cast head cylinder and model W cylinder also for adaptation to the universal test engine. In addition, it designed and built a variable fulcrum for rocker arms, a magneto mounting for the Smith radial engine, and camshaft timing adjustment for Hispano-Suiza engines. The branch designed and built intake manifolds, gun controls and magneto mounting for the Liberty 6 engine, and redesigned the Liberty 6 engine. It completed several designs for construction of air-cooled cylinder heads and different induction manifolding for the ABC, as well as the Angle indicator, and many other smaller designs. In the end, the branch considered its most important effort the model W-1 engine, an 18-cylinder liquid-cooled engine designed by McCook Field engineers in 1919, one of the largest aircraft engines attempted. By the end of 1921, the W-1 engine successfully passed its 50-hour test, and a few test engines continued evaluation into 1923.[39]

The Engineering Division designed and developed the W engine, so called because its 18 cylinders formed a "W". By 1923, the W-1 model rated 700 horsepower at 1,700 revolutions-per-minute and experts expected it to achieve 800 horsepower at 1,800 rpm without any changes. The redesigned engine was expected to produce 800 horsepower and weigh only 1,400 pounds, or 1.75 pounds per horsepower, an output comparable to the best engines then existing. A W-2 model, designed to produce 1,000 horsepower, never developed. By 1924, engine progress, notably with the Curtiss D-12, caused the Power Plant Section to conclude that the engine was obsolescent.[40]

The mission of the Dynamometer Laboratory, perhaps the largest, most modern and complete power plant testing laboratory in the U.S., was to develop and investigate all performance tests of the leading foreign and domestic aviation engines for use by the Air Service. Mr. Lake was the supervisor. The personnel of this branch, mostly experienced aviation mechanics, were among the best in the nation. The laboratory had four testing units of 400-horsepower Sprague electric dynamometers so that two units could test engines up to 1,000-horsepower capacity. Tests of spark plugs, piston rings, pistons, and related items were made on one-cylinder Liberty engines coupled to two 100-horsepower Sprague electric dynamometers. Four torque stands capable of reproducing nearly actual flight conditions were available for endurance tests. Most structural changes happened as a direct result of testing on the torque stands. The laboratory, which used more than 50,000 gallons of gasoline annually for its testing, included a large shop for overhaul of all experimental engines.[41]

In 1919, McCook Field tested a 24-cylinder X-type Liberty engine built from standard Liberty 12 parts. The changes mainly consisted of using two regular crank case upper halves, one of which was altered somewhat, and specially designed connecting rods. The motors compared:

Liberty 24
673 horsepower
1.97 lbs per hp
.55 lbs gas per hp hr

Liberty 12
400 horsepower
2.11 lbs per hp
.51 lbs gas per hp hr

An engine that ran with such power at normal speed allowed use of a comparatively large, slow-speed propeller without gear reduction, thus increasing propeller efficiency.[42]

The Installation Branch, directed by Edward T. Jones with the assistance of H.C. Osborne and W.W. Stryker, had responsibility for engine installation, control systems, fuel systems, oiling systems, superchargers, and special equipment. The branch designed, inspected, and supervised construction, and performed liaison between the Power Plant Section and the hangars. The branch described the work it performed as very professional, "like an engineering job", rather than a "heterogeneous mass of fittings." It developed engine-driven duplex sylphon pumps that proved effective devices for getting fuel to the engine. In addition, the branch developed relief valves, check valves, line connections, distributors, and related equipment. A centrifugal oil separator, one of Stryker's inventions, was in development and appeared promising. The branch also installed the Liberty 6 engine in the Fokker aircraft and the first Hispano-Suiza 300-horsepower engine in the DH-4 aircraft. The Installation Branch supported the development of specifications for engine installations used by aircraft designers.[43]

The Installation Branch was responsible for development of the Moss supercharger. The successful development of the supercharger contributed to record-breaking altitude flights of the Flight Test Section's Major Rudolph W. Schroeder, Lieutenant John A. Macready, and Lieutenant Leigh Wade.[44]

The Cooling Systems Branch was under the supervision of Lt. L.L. Snow assisted by Lt. Bayard Johnson. Workers had to carry off the heat produced by gasoline engines to avoid burning up the engines. The mission of this branch was to know all about how to dissipate surplus heat, and to apply the best methods to Air Service power plants. The branch inspected, tested, and made recommendations on cooling systems of all aircraft that the Engineering Division intended to accept. It also improved the design of cooling systems in use, designed new cooling systems, performed research, and supported standardization. The branch furnished all of the cooling system data required by the various organizations of the division and any requested by outside designers, manufacturers, and individuals.[45]

In 1920, the Cooling Systems Branch exploited a recently developed extrusion process that produced a new type of radiator core. Experiments with the core indicated effective cooling through use of high velocities from the propeller slipstream. Other advantages included ease of construction and repair, unlimited water space, a large total area with a small frontal area as well as lightness and rugged construction.[46]

The Section in 1923. When Major Hallett left the service in December 1922, Jones succeeded him as chief. During 1923, Jones restructured the Power Plant Section into six organizational elements. In addition to the chief's office and a group of Research Project Engineers, the section included four branches: Engine Design, Installations, Power Plant Laboratory, and Engine Liaison. The Chief controlled the administration of the Power Plant Section.[47]

Research Project Engineers controlled development of superchargers, carburetors, and ignitions. They also conducted extensive research work on problems related to this equipment. In addition, these engineers researched special problems involving lubrication oil and fuel exhaust gas analysis.[48]

The mission of the Engine Design Branch was to prepare the layout and detailed design of new engines and accessories, test the engines and accessories developed by the branch, and prepare reports and analyses of engine designs submitted by contractors.[49]

The Installation Branch dealt with all problems relating to installing engines in aircraft, including designing radiators, fuel pumps, oil pumps, engine controls, fuel and oil system parts, and similar equipment.[50]

The Power Plant Laboratory Branch consisted of two elements: the Engine Overhaul Unit, and Dynamometer and Torque Stand Testing. The Engine Overhaul Unit repaired and overhauled all engines used at McCook Field. This unit performed special installations, including running tests of high compression pistons or other devices. It also brought up to date all engines used in flight testing and engines used by firms developing aircraft under contract to the government. The Dynamometer and Torque Stand Unit performed calibration and testing on engines. Engineers in the Service Testing Unit concentrated on dynamometer work. These engineers calibrated the engines used by the service and determined the proper seating for the carburetors and ignition. They were also responsible for testing some experimental engines (though most such engines were tested by the engineer in charge of their development.) The Torque Stand Unit put experimental engines through 50-hour endurance testing to determine whether all major and minor parts were strong enough to withstand the severe long runs. Endurance testing included five percent at full throttle and the remainder at 90 percent power. Service engine tests took place at the torque stands to insure proper functioning. Testing of bearings and pistons also occurred before installation.[51]

The Engine Overhaul Unit tuned up practically all of the engines used in the record-breaking aircraft. Among the engine were those used by Lieutenant Macready in his altitude record, by Lieutenants Harris and Pearson in speed records over 500, 1,000, and 1,500 kilometer courses, and by Lieutenants Kelly and Macready in their endurance and transcontinental flights In addition, the unit's work supported thousands of hours of routine test flights at McCook Field with only a small number of forced landings due to engine failure.[52]

The Engine Liaison Branch kept all records related to changes in designs of recently developed engines and coordinated information between McCook Field and the manufacturers.[53]

The Power Plant Section prided itself on its close connection to the development of all successful aviation engines by 1923. Testing performed by the section led to changes that resulted in greatly increased engine life and more satisfactory performance. The section sponsored the development of new types of engines, including the most successful high-powered engine in the world. The methods used by the section in engine overhaul had become standardized throughout the Air Service and in depots repairing engines.[54]

The Power Plant Section also took initiative in solving new problems with engines. The section led the way in developing accessories for the Liberty engine, especially because manufacturers had little interest in modifying the thousands still on hand. By 1924, more than 200 changes occurred in the Liberty to improve its reliability and ease of maintenance, including air cooling. Improvements included new valves, reinforced cylinders, improved oil and fuel pumps, new carburetors, gears, improved synchronizer drive, better piston drainage, and related equipment. The tests on these parts indicated that it was possible to double the life of the Liberty engine and increase its reliability.[55]

The Power Plant Section involved itself in the development of water-cooled engines such as the Almen barrel-type, 18-cylinder engine. J.O. Almen of Seattle, Washington, had developed the Almen engine in its automotive form in his company, Almen Motors, Incorporated. The contractor constructed the engine as an aircraft engine. After testing the engine, the section regarded the engine as having great potential for pursuit aircraft because it was extremely compact, free of vibration, and required lighter mounting. Its small overall diameter and symmetrical outline lent themselves to a clean installation. In a 400-horsepower engine, for example, the motor was so compact that the minimum size of the fuselage was limited by the requirements for the pilot rather than by the size of the engine. The engine, which might have provided weight savings compared to conventional water-cooled engines, experienced failure of piston bearings in 1922 forcing redesign. Some problems with the engine never ended, and the Curtiss D-12 engine became available before resolving the difficulties with the Almen.[56]

The Curtiss water-cooled V-12, the D-12 engine of 1922, proved to be a landmark engine with worldwide influence. Based on the Curtiss K-12 engine designed in 1917 by Charles B. Kirkham, the D-12 never fully developed. Dr. Arthur Nutt, Chief Engineer at Curtiss, developed the D-12 from the K-12, a descendant of the Hispano-Suiza 8-cylinder engine. Nutt built on McCook Field's critiques in 1921 of Curtiss' interim C-12 and CD-12 engines. Subsequent test data at McCook Field helped to improve the D-12. The engine, rated at 425 horsepower, gave excellent service on 50-octane gasoline. Cylinder cooling was much better than on the Hispano-Suiza though not as good as on later engines. The Air Service used the D-12 extensively in fighter and racing aircraft. Curtiss later enlarged the D-12 into the V-1400 and then enlarged that into the V-1570. All were direct drive. Development of geared propellers and geared superchargers for this engine did not occur because air-cooled engine development took place in the U.S.[57]

Curtiss' D-12, the classic engine of this period, had a remarkable career. The engine achieved fame with successes in Curtiss racing aircraft of 1921 and 1922, and in the seaplane that won the Schneider trophy race of 1923 (considered the central achievement of the D-12, an accomplishment that led to its purchase in England.) The engine was noted for its compact arrangement and clean outline. The small frontal area allowed it to be built into a fuselage so that the minimum cross-section was established by the pilot rather than the engine. Each bank of six cylinders was cast in a block and bolted to the crankcase. It had closed-top steel cylinder sleeves screwed into aluminum-alloy cylinder heads bolted to the block casting. The pistons were of cast Y aluminum alloy. The crankcase was of cast aluminum alloy. The compression ratio was 5.3-to-1. Weight of the developed engine was 680 pounds dry; total installed weight, including radiator, piping, water, and oil, was slightly over 850 pounds. Maximum output was 440 horsepower at 2,250 rpm. A high-compression version, 6-to-1, had a maximum output of 475 horsepower at 2,250 rpm. Other modifications of the D-12 with different designations were constructed. An enlarged version of the D-12, known as the V-1400, won the Schneider race of 1925, but did not go into production. In 1926, the D-12 design led to the Conqueror, an engine of similar design that continued into development until 1933. The Conqueror developed 600 horsepower.[58]

Maintenance problems caused many disadvantages in the D-12's. Difficulty at reaching spark plugs and carburetors caused maintenance workers to remove a whole cylinder block to change a valve. McCook Field engineers considered the maintenance problems beyond the capabilities of a war-time station. But the disadvantages were offset by the engine's exceptional power, small size, and low weight. The engine achieved a virtual monopoly for fighter aircraft, and orders poured in from the Army and the Navy. Ultimately, success of the D-12 influenced development of the British Merlin. After 1925, the Curtiss company stopped development of the D-12, and the Wasp air-cooled engine quickly won support. By 1931, the D-12 was virtually forgotten except as a racing engine.[59]

The Power Plant Section also successfully developed improved carburetors. Thus, the section sponsored development of the NAU6 carburetor for the Wright H-3 engine, an improvement over the NAD6 with which contemporary aircraft were equipped. The section also designed an improved carburetor for use on the Liberty engine that eliminated many undesirable features of the Zenith US 52 carburetor, but retained its good qualities.[60]

The Power Plant Section worked to develop gear reduction and clutches, efforts pushed by Jones. A two-speed Liberty engine was developed for use with superchargers. Supercharged engines lack an adjustable-pitch propeller that held the engine speed down to a safe limit at altitudes while still permitting takeoff with the engine turning at its normal rpm. At the time, it was necessary to install an oversize propeller on supercharged engines, an approach that reduced engine speed about 500 rpm below its rated speed on the ground, thereby reducing power available for takeoff. Two-speed engines, however, allowed a supercharged engine to take off closer to its rated speed and significantly increased the performance of the aircraft near the ground. Then, as the engine speed increased to a dangerous point at higher altitude, it was possible to change gears and reduce engine speed to a safe level. A successful gear reduction developed for use with the W-1 engines that permitted the use of propellers operating at a more efficient speed.[61]

The Power Plant Section was also developing a four-engine transmission that would allow a central power plant installation in a center fuselage to deliver 1,600 horsepower. On a large bomber, this kind of transmission would allow one engine to be held in reserve for use when taking off or climbing very steeply. The engine held in reserve could also be given minor repairs while in flight. Using two or more such units in very large aircraft like the Barling bomber appeared feasible.[62]

. The Power Plant Section also was working to develop a fuel system that would be as reliable as the oil and water systems. One effort aimed at adapting the gear pump for use with gasoline to achieve a reliable fuel system. By 1923, the section developed a flexible drive expected to eliminate many of the difficulties affecting installation of fuel systems on existing aircraft.[63]

The Power Plant Section was responsible for developing the oil pump used on the Liberty engine and for developing a standard engine control unit that could be adapted to all types of aircraft. The engine control unit, designed to be practically interchangeable, was expected to have the essential characteristics of satisfactory operation and light weight.[64]

Finally, the Power Plant Section in 1923 was working on many other problems, including successful adaptation of oil heaters to all standard aircraft; the study of lubricating oils in order to determine the best for use in aircraft; the study of rotary induction systems to improve distribution in aviation engines; and the study and compilation of data relating to all contemporary types of radiators, including the use of ethylene glycol (later called Prestone in its trade name) as a coolant. Testing of ethylene-glycol as a high-temperature cooling fluid was first experimented with by Heron in 1923 on McCook Field's 1-cylinder engine. In the late 1920s, further testing at McCook Field on a D-12 engine was followed by development to overcome leaks and other problems. Eventually, ethylene-glycol resulted in the gradual obsolescence of water-cooling engines in favor of liquid-cooling, marked reduction in the size of aircraft radiators -- about 70 percent -- and significant savings in weight and drag. Thus, the drag of liquid-cooled engines fell below air-cooled radials of equal power, and their installed weight also came down to more comparable levels. Air-cooled fighters, however, proved better at lower altitudes because of lighter weight and less vulnerability to small-arms fire.[65]

. In 1924, the Power Plant Section had 94 of the Engineering Division's 1,087 employees and a payroll of $199,003.27. In 1925, the section had only 91 of the division's 996 employees and a payroll of $194,124.27. At that time, the Power Plant Section's mission was summarized in these words:[66]

> *The functions of the Power Plants Section are as follows: Research, development, design, construction, and test of airplane power plants, special equipment and accessories, including superchargers, carburetors, ignition systems; cooling systems, lubrication and fuel systems; maintenance of a liaison with the contracting aircraft*

engine industry and Air Service at large; and the overhaul of special service engines for test, stock, and contracts.

End of an Era: 1926. In October 1926, the Materiel Division was established with temporary headquarters at McCook Field. In this reorganization, the Engineering Division fell an echelon lower in the organizational hierarchy, becoming the Experimental Engineering Section of the new division. Major Leslie MacDill continued as chief of engineering, a position he assumed in 1923. Major John F. Curry, commandant of McCook Field since July 1, 1924, retained that position until August 1927. As part of the reorganization, the Power Plant Section of the Engineering Division also fell an echelon lower, becoming the Power Plant Branch of the Experimental Engineering Section. Likewise, the Airplane, Armament, Equipment, Lighter-Than-Air, and Material Sections of the former Engineering Division were reduced to branches in the new division.[67]

When the Materiel Division formed, construction of aircraft and accessories had already turned over to aircraft manufacturers and related industries, leaving the division to formulate types and to perform research and experimentation that industry could not do. The purpose behind this policy, mandated by congressional legislation in 1926, was to help build up and stabilize the aviation industry and afford it an advantageous position for the manufacturing of commercial aircraft. In 1926, the biplane was still the most common type of construction though monoplanes were under order by the Materiel Division. Metals, including steel, chrome-molybdenum, duralumin, and magnesium, were also rapidly replacing wood in aircraft construction.[68]

In 1926 and 1927, the Power Plant Branch continued its research and development in propulsion. The essential thrust in engine development was toward more power at higher engine speed but with less weight per horsepower, a movement that favored air-cooled engines. However, the branch did not "feel safe in entirely abandoning the development of water-cooled engines." Thus, work continued on Curtiss and Packard engines. One new water-cooled engine was the Curtiss 12-cylinder, 600-horsepower V-1550 Conqueror, a slightly larger version of that company's D-12 that was in quantity production for the Army and Navy. Compared to the direct-drive model of Curtiss' earlier engine, displacement of the new engine was 1,550 cubic inches versus 1,145; 600 horsepower versus 435 horsepower at 2,300 rpm; and 1.33 pounds per horsepower rather than 1.84 pounds. The V-1550 weighed only 730 pounds dry. Another 12-cylinder V-type, the Wright Tornado (T-3), offered 600 horsepower. Packard continued developing its 1A-2500 and 1A-1500 types. The company's new 12-cylinder V-type engine, the 2A-1500 series, incorporated a new crankcase design along with other features, but weighed slightly more than the Curtiss 1550. The rated output of this Packard engine increased to 600 horsepower at 2,500 rpm. It was produced in three types, direct-drive, geared, and inverted.[69]

Packard's 800-horsepower 1A-2500-geared engine was in service testing on the LB-1 single-engine bomber. Development of that V-type 12-cylinder engine included improved cylinders, valve housings, ignition system, and other features from the 2A series. Additionally, a radical change in design was to add a gear-driven supercharger on both sides of the crankcase; this special model, the 4A-2500, was expected to develop 1,000 horsepower at 2,300 rpm.[70]

Packard recognized that in developing its 800-horsepower 2500-series engines McCook Field's propulsion experts played a key role. The company's test report noted that[71]

> *a considerable share of the credit belongs to the Engineering Division of the Air Service who not only displayed exceptional far sightedness in laying down the original program but also cooperated with helpful suggestions during various phases of the development.*

This testimonial echoed the praise of other contractors who appreciated the division's valuable engineering assistance in solving their problems.

Despite the rapid development of new engines, Liberty engines were still available by the thousands and in widespread use in 1926 and often competed directly when new aircraft were designed. Not until 1934 did the Army stop using Liberty engines. Liberty engines were available in both standard and air-cooled types. Frederick W. Heckert in the McCook Field shops successfully inverted the Liberty in 1923. After the fuel and water flow problems were resolved, the inverted Liberty improved visibility, maneuverability, and accessibility for maintenance of the standard DH-4 aircraft in which it was tested. Grover Loening's successful amphibian, designated the XCOA-1 aircraft by the Army, used the inverted Liberty most widely from 1924 to 1930.[72]

Liberty engines still challenged more modern engines like the Curtiss D-12, V-1400, and 9-cylinder R-1454 (an air-cooled engine using the Engineering Division's type M cylinder) for use in pursuit aircraft. Four Liberty engines, each pair driving a 17 1/2-foot propeller through a special transmission, imparted a top speed of 80 miles per hour to the Army's new RS-1 airship, the first semi-rigid built in the U.S. Development of the RS-1 required four years of effort in cooperation with the Goodyear Tire and Rubber Company.[73]

Propulsion equipment in development by the Engineering Division included improved bearings, spark plugs, carburetors, fuel metering systems, lubricating oil, and cooling systems. Thus, the dangerous method of starting aircraft engines by swinging the propeller rapidly disappeared as reliable, compact, light starters of both mechanical and electrical types developed. Electric engine starters could crank engines with as much as 1,000 horsepower, and could do so quickly and safely. Work was being done on ignition devices such as the new double-type Scintilla Magneto Company and Splitdorf Electrical Company magnetos that combined in a single unit two independent breakers driven by one rotor, thereby allowing a 50-percent reduction in weight by eliminating two magnetos formerly required by an engine. Engineers also developed a new type of pivotless breaker that increased magneto endurance from 100 hours to more than a thousand hours of continuous operation without adjustment. In addition, they developed a refueling pump capable of delivering 1,000 gallons an hour. Because the pump weighed only 15 pounds, it could be carried as portable equipment on aircraft. Other equipment in development included batteries used for ignition and starting, and engine-driven generators for charging the batteries. A four-engine Liberty transmission that could drive one propeller was developed based on Power Plant Section specifications. The transmission consisted of a large rectangular cast aluminum case with four driving pinions grouped around a large spur gear on the propeller shaft. Each pinion was driven by a Liberty engine through a sliding tooth clutch so that any engine could be thrown in or out of gear at will. Shifting gears was as easy as in an automobile.[74]

Among the achievements listed by the Engineering Division in 1925 was the design and construction of the first detachable power plant unit long before such an arrangement was used abroad. The power nacelles on the G.A.X. aircraft were demounted by removing four bolts and disconnecting the engine controls, a design used in constructing the aircraft. Other achievements were the first droppable gasoline tank enclosed in the fuselage and the method of guiding the tank after it was released. This tank was used on the PW-1 pursuit.[75]

Another achievement of the division was publication of the Handbook of Instructions for Airplane Designers. This book listed detailed requirements for structures, power plant installations, and equipment based on the experience of the division and the Air Service. The data in the book was especially valuable since it was compiled from various reports that were not available to the individual designer. Thus, the division considered this publication a direct aid to industry.[76]

Technical Thrusts

In addition to the technological accomplishments discussed in connection with the organizational history of the Power Plant Section, the section participated in several breakthroughs in other key areas. Among them were air-cooled engine technology, superchargers, and fuels.

Air-Cooled Engines. The Power Plant Section described its role in the development of air-cooled engines as antedating both contractor and Navy interest. In 1920, propulsion contractors dismissed the possibility of effectively air-cooling engines, arguing that the development of a large air-cooled engine was neither feasible nor desirable. Moreover, the Navy's Bureau of Aeronautics went so far as to state that such an engine would be of no interest to them, even if developed, and refused to take a share in this preliminary work. By 1924, of course, those attitudes changed dramatically, and the division pointed to its pivotal role in this transformation:[77]

> This whole situation is due absolutely to the fact that the Engineering Division, working unaided, succeeded in proving conclusively that large air-cooled cylinders were not only entirely feasible but could be made equal to the best water-cooled cylinders, both as to efficiency and reliability.

To substantiate this claim, the division quoted from an article published in 1924 by an engine expert from the Navy's Bureau of Aeronautics:[78]

The Engineering Division of the Army Air Service has in its excellent research work at McCook Field demonstrated the entire practicability of constructing air-cooled cylinders of large displacement that have thermodynamic characteristics equal to the best water-cooled construction. It remains only to solve the relatively simple mechanical details of combining such cylinders into a large-power engine without exceeding reasonable frontal dimensions.

Support for the Engineering Division's case was provided a half century later by historian Walter J. Boyne in his analysis of McCook Field's contributions to aviation. Pointing to Lawrance's J-1 engine as fundamental, he added:[79]

From this basic engine emerged the famous Wright J-4s, J-5s, and in a very real sense, all other radials by Wright [Aeronautical Corporation] and Pratt & Whitney. It's probably not stretching the truth too much to say that the Engineering Division's foresight in backing Lawrance laid the foundation for air supremacy in World War II when Pratt & Whitney alone turned out almost 370,000 round engines. [Boyne's italics.]

Based on experience in the Great War, water-cooled engines had gained superiority over air-cooled engines because of their greater reliability and compactness. Even into the 1920s, water-cooled engines offered higher speed and greater power. Late in WWI, the Royal Aircraft Establishment in England began research on air-cooled cylinders. In 1915 and 1916, Dr. A.H. Gibson and Heron performed exhaustive analyses of the cooling requirements and the type of construction most suited to air-cooled cylinders. After the war, the Power Plant Section concluded that U.S. development of air-cooled engines was trailing European progress. But when Heron migrated to the U.S. in 1921, investigations at McCook Field accelerated. Heron's work for the section's Engine Design Branch eventually led to his patent on salt-cooled valves for aircraft engines. These valves were rapidly applied to large engines, both air- and water-cooled. Salt cooling, almost as effective as sodium cooling, was subsequently abandoned because rapid starting of an engine from cold sometimes caused the valve stems to swell and stick. In 1928, Heron adopted liquid sodium as the internal coolant for hollow valves. Heron's work on the internally cooled exhaust valve permitted improvements to be made in air-cooled engines for more than 20 years, and in WWII was used in almost every tactical aircraft flown by the AAF and the Navy.[80]

The Power Plant Section supported development of air-cooled engines for a number of reasons. Because the radiator and cooling system were not needed, air-cooling offered greater reliability. Air-cooled engines saved about 0.6 pounds per horsepower in installed weight compared to water-cooled engines; thus, the installed weight of a 400-horsepower engine was about 200 pounds lighter than a water-cooled engine of comparable power. Air-cooled engines were less vulnerable to machine gun fire, and they offered increased performance because of less weight. The work of the Power Plant Section in developing in its own laboratories and shops the first large air-cooled engine cylinders successfully operated in this country formed the basis of all air-cooled engine development in the U.S. At McCook Field, engineers examined and tested cylinder-head designs, valve arrangements, valve materials and cooling methods, cylinder head attachments, spark-plug locations, cooling efficiency, valve mechanisms, and a hundred other items affecting the design of the air-cooled cylinder. Heron's work on cylinders for air-cooled engines led to a reliable type of cylinder for incorporation into new designs of air-cooled radial and V-type engines. A cylinder type was developed that surpassed the performance characteristics and durability of the best water-cooled engines. The cast aluminum-alloy head screwed and shrunk on a machined forged-steel barrel became the accepted type for high-performance air-cooled engines. Using aluminum as a cylinder material (all cylinders had steel liners) provided real advance in the performance of air-cooled engines.[81]

Meanwhile, in 1916, Charles Lanier Lawrance, influenced by the Clerget 3-cylinder, 60-horsepower engine, started developing air-cooled engines of small power. He formed the Lawrance Aero Engine Company in 1917, producing 2-cylinder A and N models and a 3-cylinder B model for the Navy. In 1918, Lawrance produced an L-series 3-cylinder, 60-horsepower radial weighing less than 150 pounds for the Navy. The Engineering Division tested the L model, and purchased several of the engines in 1921 for its Sperry Messenger liaison aircraft. The division also used a Lawrance 3-cylinder engine on its blimp. In 1919, the Engineering Division gave Lawrance a contract for a 9-cylinder engine, model R. Following extensive tests of this model, and modifications suggested by McCook Field engineers, the division awarded a contract dated 25 March 1920 for three more R engines. When delivered in 1921, the new Lawrance engines passed a 50-hour test, developing 147 horsepower at 1,600 rpm with a weight of 410 pounds.[82]

Development of air-cooled engines, especially favored by both the Navy and commercial interests, rapidly progressed in the 1920s. The Navy contracted on February 28, 1920 with Lawrance (the company consolidated with Wright Aeronautical Corporation in mid-1923 under pressure from the Navy's Bureau of Aeronautics) for a slightly larger engine, the J-1, with 200 horsepower at 2,000 rpm. This engine passed its 50-hour acceptance test early in 1922, becoming the forerunner of the famous Wright Whirlwind J-series (essentially the Lawrance J engine with the screwed-head cylinder.) The 200-horsepower air-cooled Wright Whirlwind R-790 engine, with a dry weight of 508 pounds, became a popular engine in naval and commercial aircraft after it went into production in 1925.[83]

In 1921, the Navy made a commitment to air cooling as simpler, lighter, less vulnerable, more compact, easier to manufacture, and easier to maintain and overhaul -- factors important for carrier aircraft. Air-cooled engines dispensed with the water-cooled engine's plumbing, considered dead weight, and allowed converting that weight into payload and a smaller aircraft. The Navy's experience with air-cooled engines, and its arguments for using their weight advantage over outdated Liberty and modern water-cooled engines, were summarized in an oral report to Admiral Moffett:[84]

I've looked up the records and found that it costs about a thousand dollars to convert an old Liberty into one incorporating all the new changes. After that we can get 75 hours flying time out of it before we have to put in another converted one. For 300 hours flying time, we spend four thousand dollars on conversions. Meanwhile we can get a new Wright Whirlwind for about the same money and run it for 300 hours without overhaul.

In air-cooled radial engines of 400 horsepower, the only development in the U.S. in 1923 was the type R-1 engine. The R-1 was originally built in 1920 by Wright Aeronautical Corporation after a competition held by the Engineering Division in 1919. The R-1, which incorporated the type J cylinder developed by Heron, was tested in 1921. Although the engine proved less than satisfactory because some major parts failed, Wright Aeronautical undertook a redesign. Had the engine been successful, it was expected to develop 410 horsepower at 1,700 rpm and weigh only 720 pounds, that is, about 1.76 pounds per horsepower. But in a price competition, Curtiss won the contract to develop the R-2, later designated the R-1454.[85]

Three model R-2, 400-horsepower, air-cooled radial engines were under construction by the Curtiss company. The Engineering Division recognized that the engine would necessarily carry the contractor's name and that the division's contributions would not be identifiable. But the engineering data turned over to the contractor by the division included complete cylinder design, including valve gear and piston. The cylinder, developed by the Engineering Division, was tested more severely than any water-cooled cylinder. The division also furnished the manufacturer with valve springs, cylinder castings, piston castings, and some materials that were difficult to obtain in small quantities. The division gave the contractor a complete design of a supercharging induction system based on more than two and a half years of research by the division. The division paid General Electric Company $3,000 to check and refine the design, and manufacture certain parts for Curtiss. The division provided complete drawings of the R-1 engine, including written discussion of all testing conducted on the engine and a complete weight analysis and an assessment of the design's good and bad points.[86]

In addition, the division turned over to Curtiss the complete drawings of the Allison Engineering Company's design submitted in competition with Curtiss. The division paid Allison $3,000 for the rights to their design, which included many ideas and suggestions not in the Curtiss design. Similarly, the division gave Curtiss all of the data obtained from endurance testing of the Bristol Aeroplane Company's Jupiter 9-cylinder air-cooled radial and the Armstrong-Siddeley Jaguar engine (both in development by 1917) at a time when no other organization in the U.S. had such information. These engines were completely disassembled and the parts laid out for Curtiss designers to study.[87]

Finally, the division provided Curtiss with complete drawings of an experimental lubrication system for radial engines, designed lubricate the main bearings and at the same time prevent excessive oiling of the cylinders. The division experimented with this device as a means of reducing the excessive oil consumption of the Lawrance J-1 engine. The results were very satisfactory, and Curtiss incorporated the system into their new engine.[88]

Schlaifer cited this R-1454 development, the Army's first high-power air-cooled radial, as a project with "complete Army dictation of all the basic principles of the design." But the R-1454 was slow to emerge after 1923, and was dropped after P&W's Wasp demonstrated decisive superiority in 1926. Schlaifer attributed the slowness not only to Curtiss' peculiar business position -- the company's D-12 was an excellent engine and design rights for the R-1454 belonged to the Army -- but also to the split between construction by the contractor and design and testing by McCook Field. Successful contractors were able to focus on a few projects but Power Plant Section engineers dealt with many projects.[89]

Thus, despite the Power Plant Section's support for the R-1454 engine, the Curtiss company favored water-cooled engines where its D-12 engine had an advantage and did not produce the R-1454. In 1926, when the contract was canceled by Captain Tillinghast, the R-1454 weighed 830 pounds compared to the Wasp's 650 pounds.[90]

In addition to the radial engine, two other cylinder arrangements in air-cooled engines were explored: the 400-horsepower V-1410 Liberty, a V-type in both upright and inverted models, and the 1,200-horsepower Allison Engineering Company X-4520 24-cylinder X-type model based on the Liberty. The Engineering Division continually modified Liberty engines after WWI. It also contracted for air-cooling the Liberty to demonstrate the practicality of an air-cooled V-type engine. Heron designed the air-cooled version and Allison Engineering Company performed the work on the engine, designated the V-1410. The air-cooled version provided a decreased frontal area compared to the water-cooled model of 3.1 square feet versus 8.7 square feet. The cylinder construction was a modification of the type J cylinder used in large air-cooled engines. The exhaust valves were salt-cooled. The horsepower of the air-cooled Liberty was about the same as the water-cooled version and fuel consumption remained about the same. In addition to a decreased frontal area, especially on the inverted engine which permitted a narrow streamline nose cowl, the air-cooled model provided weight saving of 141 pounds. The air-cooled inverted engine was also simpler to install, more accessible, and quieter. But the air-cooled Liberty proved too heavy to compete with the radials then in development.[91]

Another engine designed at McCook Field was Harold Caminez's cam model which did not use a conventional crankshaft. Caminez designed this 4-cylinder, X-type, radial, air-cooled engine while he was in charge of engine design for the Engineering Division. An aircraft engine was adapted for use with the cam drive and the principles of the device tested in the Power Plant Section's laboratory. Sherman M. Fairchild formed the Fairchild-Caminez Engine Corporation in January 1925 as a subsidiary of the Fairchild Aviation Corporation to build cam-type aircraft engines. The first engine using the cam mechanism was completed by the company in 1925 and given a 50-hour test at McCook Field. First flight with the engine was made in the company's aircraft in April 1926. The model 447-B engine was the outgrowth of this development. The engine had 447 cubic inches of piston displacement and developed 142 horsepower at 1,120 rpm. Plans for producing the engine went forward in mid-1927. Ultimately, the engine proved impractical because of excessive vibration resulting from torque variation.[92]

Meanwhile, Pratt and Whitney Aircraft Company was formed in August 1925 as a subsidiary of the Pratt and Whitney Company, established as a manufacturer of precision tools and machinery in 1860. The men behind the formation of P&W were Frederick B. Rentschler, who left as chairman of Wright Aeronautical in 1924; George J. Mead, a Wright engineer and earlier chief of the Power Plant Laboratory at McCook Field; A.V.D. Willgoos, another engineer; and some others. During 1926, P&W turned out a 9-cylinder 425-horsepower radial, the Wasp, for Navy pursuit and observation aircraft. The engine weighed only 650 pounds, a savings of 200 pounds compared to the D-12. The Wasp, developed almost as fast as the original Liberty engine though it was not produced as quickly, was the first successful large radial air-cooled engine of modern design. The engine was also used commercially and proved its capability in service. Although the Wasp was "probably the best unsupercharged direct-drive air-cooled engine in production anywhere in 1927," according to Schlaifer, even larger radials appeared in 1927: Pratt and Whitney's 9-cylinder 525-horsepower 750-pound Hornet R-1690 (a larger version of the Wasp), and Wright Aeronautical Corporation's 9-cylinder 525-horsepower Cyclone R-1750 engines (both geared and direct-drive) for larger aircraft. The Cyclone engine was designed by Lawrance, Jones, Heron, and Morehouse after the latter three joined Wright Aeronautical in 1926. The R-1820 Cyclone went into production in 1927, the year that Lawrance won the Collier trophy for the radial air-cooled engine.[93]

By 1925, the Engineering Division had developed an enamel for application to air-cooled engines. A glossy smooth finish on cast surfaces was found to increase heat dissipation, in the case of a coating on cast aluminum cylinders, by as much as 15 percent. Service conditions required an enamel that would not check or burn off at temperatures up to 700 degrees Fahrenheit and was elastic enough to withstand vibration and the strains of contraction and expansion. A mixture of castor oil, Gilsonite gum, and mineral spirits was baked on the metal at about 400 degrees Fahrenheit. Tests indicated that the enamel was entirely satisfactory, and the formula was used by a varnish manufacturer to make production batches.[94]

In conjunction with the development of air-cooled radials, the Power Plant Section also was studying the design of a cowling to reduce the drag caused by their larger frontal area. A successful cowl would enable air-cooled radials to compete with water-cooled engines in head resistance (except where wing skin radiators were used.) However, not until 1929 was the drag greatly reduced, an improvement brought about by effective cowling and cylinder baffling developed at Langley Field by NACA. The NACA cowl permitted much better airflow and less turbulence.[95] As early as 1926, air-cooled engines were beginning to challenge in the National Air Races. In 1928, the National Air Races were won for the first time by an air-cooled engine. All of the entries were powered by Wasp engines, but the winning speed was only 172 miles per hour (compared to 206 in the 1922 Pulitzer and almost 215 in 1925.) Air-cooled engines powered the winners of all major national races during the decade from 1931 to 1940.[96]

105

Supercharges. Aircraft engines lost much of their power and range at high altitudes where the atmosphere was thinner. In 1914, a Swiss engineer, A.J. Buchi, suggested the possibility of applying a turbo-supercharger to aircraft. Dr. Auguste Rateau in France built a turbo-supercharger in 1918 that was driven by the engine exhaust through a turbine. But further work was needed to make the device practical. The armistice occurred before the supercharger was ready, however, and none was put into service in WWI. French and British work on exhaust-driven superchargers was eventually discontinued in favor of gear-driven types.

In 1917, Dr. W.F. Durand of NACA, aware of General Electric Company's work with gas turbines and centrifugal compressors, requested the company's cooperation in developing a turbo-supercharger to maintain the power of airplane engines during the air battles at high altitude in WWI. During the war, aircraft operated at altitudes up to 20,000 feet where difficulties with carburetion and distribution, ignition and plugs, liquid cooling, and propellers were encountered. The Bureau of Standards did theoretical work, but McCook Field engineers did practical work in cooperation with Mr. E.H. Sherbondy of the DeLaval Company and, subsequently, with Dr. Sanford A. Moss of GE. Dr. Moss's supercharger, the exhaust-driven type, was the one eventually finally brought to the practical stage.[97]

The Engineering Division contributed significantly to development of the supercharger. Developing the device required familiarity with an area of engineering not previously associated with aircraft power plants. Therefore, engine manufacturers possessed little experience in this area. Virtually the entire development resulted from cooperation between the Engineering Division and General Electric. Because supercharger development consisted essentially of producing a high-speed exhaust-driven turbine air pump, GE was the best qualified U.S. manufacturer. The company's engineers had data relative to superchargers that was generally unavailable, but they lacked knowledge of aeronautical work and facilities for flight testing. Moreover, the form and type of construction used in the pump had to be suitable for installation and operation in military aircraft, another area where GE lacked experience. The Engineering Division concluded that it was easier to train their propulsion experts in the essentials of turbine pump design than it was to familiarize GE engineers with the requirements of aircraft service.[98] From 1917 to 1921, the supercharger underwent vigorous development by Dr. Moss and the Power Plant Section. Experimental models were tested on a Liberty engine at the top of Pike's Peak (14,109 feet) in September and October 1918, and in flight at McCook Field in 1919. An unsupercharged Liberty engine could develop 400 horsepower at sea level, but only 200 horsepower at 15,000 feet and only 87 at 25,000 feet. In September 1921, Lieutenant Macready took a LePere biplane, P-53, with a supercharged engine to a true altitude over 34,000 feet. Without the supercharger, the LePere could not rise above 20,000 feet.

The function of the supercharger was to "pep up" the engine, enabling an aircraft to operate effectively at high altitudes by compressing air drawn in from the atmosphere even where the air thinned out. For example, the LePere had a service ceiling of 20,200 feet with a speed of 92 miles per hour; but with a supercharger, the aircraft could exceed 30,000 feet and could reach 142 miles per hour at the service

106

ceiling of the unsupercharged aircraft. An unsupercharged DH-4B aircraft had a service ceiling of 15,000 feet and a speed of 103 miles per hour; but with a supercharger, the DH-4B attained a speed of 127 miles per hour at 15,000 feet, 133 mph at 20,000 feet, and had a service ceiling of 27,000 feet.[99]

The supercharger developed by Dr. Moss was a high-speed centrifugal air pump driven by an exhaust-gas turbine. Some power was still available after the charge in the engine cylinder was operated. The supercharger turbine converted this power into useful work, driving the compressor that pumped air to the engine at constant sea level pressure, thereby maintaining its full power. The first and most serious difficulty was obtaining a material that could withstand high temperatures from the white hot exhaust gas. The nozzle box and manifolds got sizzling hot, and the turbine wheel spun at over 22,000 revolutions per minute in gases at 1,500 degrees Fahrenheit. One by one, however, the difficulties were solved. Flexible manifold connections helped to overcome warped and cracked joints and leaky connections. The tips of the turbine wheels were ground off to lighten them and to decrease the stress of centrifugal force.

A side-type supercharger followed, inspired by Jones, chief of the Power Plant Section. In Jones' arrangement, the turbine shaft was placed at right angles to the engine crankshaft, the turbine wheel was in the open -- overhung on the nacelle surface and exposed to the full blast of the propeller slipstream to keep the turbine buckets from overheating. The turbine nozzle box was flush with the cowling, and the compressor was enclosed by the cowling. By mounting the compressor casing sideways, head resistance was cut down. The whole installation was simple, easily adjusted, and worked. The first unit ran for over 50 hours without a problem. Subsequently, the side-type supercharger was subjected to numerous flight tests, and demonstrated significant advantages. By 1924, this type of supercharger was standardized as reliable, rugged, and easily maintained. Experiments with superchargers at altitude were included in the nearly million miles flown by pilots at McCook Field during 1926. Although application of superchargers in commercial aviation at the time was lagging, the basic design developed by the Power Plant Section was used in racing cars, rail cars, and the gasoline-electric locomotive. The design proved effective and was used on nearly all turbo-supercharged aircraft through World War II.

A special unit, the 35,000-foot supercharger, was designed for work at exceptionally high altitudes. It resembled the Form B supercharger, but the speed of the turbine was increased from 22,000 rpm to 34,000 rpm. The turbine and impeller of this model were given an overspeed run at 41,000 rpm. The tip speed of the impeller reached 1,880 feet per second, three-fourths the speed of a rifle bullet. The centrifugal pull on each bucket weighing .00959 pounds, was 1,750 pounds. Thus, the stress was very high. The turbine and shaft were machined from a solid forged mass of steel weighing 250 pounds. The finished weight of the turbine was only 20 pounds, eight percent of the original forging. A commercial compressor delivering the same amount of air as this supercharger would have weighed 5,000 pounds and taken up a space six feet by six feet by eight feet. At 35,000 feet, the supercharger increased the power of the Liberty engine by over 300 horsepower.[100]

Because exhaust-driven turbo-superchargers caused difficulty in cooling valves and reduced power somewhat, development of superchargers also focused on a gear-driven model. The principal problem with this type was developing a drive coupling that could withstand sudden acceleration of the engine. Two types of gear-driven superchargers were in development. One type used air that did not pass through the carburetor and used a centrifugal fan to compress the air as it entered the engine. Changing gears in this type made it possible to give ground atmospheric pressure at high altitudes. This type was developed with the support of the Engineering Division. The second type of gear-driven supercharger, the Roots type, compressed air before it reached the carburetor. NACA and the Navy were involved in developing this type.[101]

Work by the Power Plant Section to solve the problem of gear-driven superchargers was another initiative of Jones. The section believed that practically all the European nations regarded the task as impossible. But the section felt that successful gear-driven superchargers were possible, and that application to pursuit aircraft would give the U.S. superiority in maneuverability and speeds at high altitude. Geared superchargers were built for supercharging at 7,500 feet and at 20,000 feet. The most difficult part of the 20,000-foot supercharger, the development of satisfactory gears, appeared solved by 1925 since the gears had stood up very well in all cases. Additionally, the division noted that development of a 7,500-foot geared supercharger had the potential to improve the performance of engines to such an extent that it might have to be incorporated on all types. The Power Plant Section also supported development of an automatic control for superchargers.[102]

The experimental superchargers in construction by GE in 1925 were based on preliminary designs prepared by the division. The division laid out the general form of the device and the type of construction before the company's engineers became involved. GE then checked the design from their viewpoint and suggested changes for approval by the division. When the design was satisfactory to both groups, the division placed an order for construction. According to the division,[103]

> It is thus apparent that the Engineering Division does the major portion of the design work on the General Electric superchargers, yet, when the devices are developed and reach the service they are invariably referred to as General Electric superchargers and this Company is credited with the entire development work.

Although the Power Plant Section felt that it would not receive proper credit for its contributions to the development of superchargers, the role it played was hailed by one knowledgeable observer:[104]

> The turbosupercharger is an extremely valuable contribution which is due financially to the Army's experimental funds alone, and technically to close collaboration between the Army and General Electric. No engine builder showed any enthusiasm for the turbo until the 1940's, let alone assisting financially in its development, and no other example shows so clearly that the military services have on occasion been fully justified in

108

insisting on a particular technical line although the engine builders were uninterested if not hostile.

Fuels. Improved fuels were another area for research and development at McCook Field. As early as 1917, Major Henry Souther arranged with the Bureau of Mines to study the problems associated with aviation fuel. Gasoline originating in different parts of the U.S. were compared, including California, Oklahoma, and Pennsylvania. Tests at Langley Field indicated that engines operated better with volatile fuel than with heavier types of gasoline. Meanwhile, in Dayton in 1917, Charles F. Kettering promoted the study of anti-knock substances, leading to the cooperation of the Dayton Metal Products Company and the Bureau of Mines with the Dayton Engineering Laboratories Company (Delco.) Test flights were also made at both the Dayton Wright Airplane Company and Wilbur Wright Field.[105]

Thomas Midgley Jr. of McCook Field ran tests on a 1-cylinder engine, bringing promising substances to the field's propulsion laboratory for testing in an aircraft engine from 1919 to 1923. Toluene compounds appeared promising by 1920, and metallo-organic compounds in 1921. In 1922, Midgley brought the first sample of tetraethyl lead to McCook Field for tests in 1-cylinder and full-scale aircraft engines, and experimental work with leaded fuel advanced rapidly after that. The Navy adopted lead for aviation gasoline in 1926, and the Army in 1933. Meanwhile, by 1925, the Power Plant Section had developed a complete line of standardized fuel system and installation parts and fittings, including several hundred individual parts.[106]

The Propeller Branch

Frank W. Caldwell was chief of the Propeller Branch in the Airplane Section from 1918 to 1930. Caldwell came to work for the Army from Curtiss.

Progress Through 1927. Within the Shop Engineering Section at McCook Field was a Propeller Sub Unit supervised by Caldwell, the engineer representing the Airplane Section. The sub unit was directly under the supervision of R.J. Meyers, general foreman. In 1919, the unit designed 120 propellers and in 1920, another 50. After testing, designs were standardized for production aircraft, including the USD-9, the Glenn Martin bomber, LePere, VE-7, and the SE-5. Other propellers were furnished for experimental aircraft. The unit also developed variable- and reversible-pitch propellers, and metal and micarta propellers. The Propeller Test Laboratory at McCook Field used motion pictures in its wind tunnel experiments to visualize flight vortices at speeds up to 500 miles per hour. The Propeller Sub Unit's work resulted in the publication by the Engineering Division of a Manual on Propellers with information on design, testing, storage, and care of propellers.[107]

Wood propellers were made from birch, walnut, oak, mahogany, and cherry. They were covered with Irish linen or shoe lining to protect the leading edges and tips from wear and damage caused by rain and small stones. The linen was put on with glue and ironed smooth. Propellers were inspected for length, shape, and size, tested to see that they tracked, were balanced, and were perfect in angles before being turned over for varnishing. The propellers were doped with a clear varnish that caused the linen to shrink and become very tight. High-grade varnish was used as sizing and allowed them to dry before applying an aluminum coating. Experiments on wooden propellers led to a method of coating them with aluminum leaf to prevent their absorbing moisture and to reduce damage during storage. The aluminum was put on and rubbed with a soft cloth until it became very smooth.[108]

Propellers operated under tremendous stresses. The efficiency of engines depended on many variables. It was difficult to design a propeller that could be efficient under all conditions -- takeoff, climb, maximum speed, and cruising speed. Thus, propellers in the early 1920s represented compromises designed to give the greatest efficiency during average flying conditions of an aircraft. But compromises restricted the aircraft's flexibility. Making a propeller with a variable pitch (the angle at which the blade is set), preferably adjustable by the pilot in flight, gave the aircraft greater flexibility. Variable-pitch propeller mechanisms had to be light, strong, and reliable. Reversing the propeller and controlling its operation in reverse enabled aircraft to land more safely. Experiments with these types were underway in 1919. Test results were encouraging by late 1926.[109]

The reversible propeller was developed from a design by Seth Hart of Los Angeles, California, and by late 1919, the Engineering Division had made several flights with the propeller. The mechanism of the propeller consisted of two separate blades with ferrules in the end nearest the axis. The ferrules turned in a hollow cylindrical hub barrel. As the ferrules turned in the barrel, the pitch angles of the two blades also changed until the propeller was changed from a right-hand screw to a left-hand screw. Since the engine continued to turn in the same direction, the propeller became a pusher instead of a tractor and functioned as a powerful brake.[110]

The Propeller Sub Unit also experimented with metal and micarta propellers by 1919. As early as June 1917, Westinghouse Company officials began cooperating with the government in applying micarta products to aircraft, working closely with Caldwell. Micarta propellers were made from a composition of canvas or paper impregnated with bakelite or other phenolic condensation products and molded under heat and pressure. Micarta offered definite advantages for propellers compared to wood. Tests by Westinghouse and subsequently at McCook Field showed that micarta improved climb and speed. The Propeller Branch concluded that micarta had greater efficiency, greater strength, greater resistance to wear when striking grass, sand, and small stones, much greater resistance to rain and hail, less noise, and many times longer life. Micarta was not affected by climatic conditions. Westinghouse received a contract to produce several micarta propellers, and the company built manufacturing equipment. Although micarta proved highly effective, it was also relatively costly.[111]

By 1923, the Propeller Branch of the Airplane Section achieved considerable sophistication with its work. The branch was designing its propellers based on theory adjusted for empirical data from flight tests. McCook Field exploited the two-bladed propeller, laminating it using boards three-quarter inch thick. The boards were cut from carefully selected lumber with a uniform texture and with high tensile and shearing strength. Preferred pieces were free of all defects and had a straight grain running the entire length. The wood was then kiln dried. After cutting to shape, the boards were individually balanced and selected for grouping together for the complete propeller. The net amount of wood present in a two-blade propeller varied from about 30 board-feet in training types to 80 in combat types. The gross lumber required to manufacture these propellers was about twice the net amount.[112]

After the propeller was finished in the raw wood, it was inspected and then finished with five coats of spar varnish. The propeller was then rebalanced until no consistent motion was shown in any direction. All wooden propeller tips were covered with pig skin, cloth, rubber, Monel metal, or terne plate (the Air Service used it frequently.) Absence of any roughness in the blade was necessary since the tip speed for some types was a thousand feet per second.[113]

One major function of the Propeller Branch was testing propellers, both for McCook Field and for outside firms (for which charges were assessed based on time and horsepower.) Testing was done on a destructive whirling test rig to reduce the risk that a propeller would break in service on an aircraft because the unbalanced forces resulting from a broken propeller could wreck the aircraft. Propellers were tested for 10 hours at either twice the horsepower or up to twice the rotation speed at which they would have to operate. Water spray tests could also be made. After successful testing on the rig, propellers were flight tested on aircraft for which they were designed.[114]

Charles Fayette Taylor described one of the misadventures that he and Caldwell experienced while testing a metal propeller at McCook Field:[115]

> In 1921 Caldwell tested a steel-bladed propeller on his electric whirling machine to twice its rated power. He then, very innocently, presented it to me for a "routine" test on a Hispano-Suiza 300-hp engine. After a few minutes at rated power, a blade broke off, came through the control board between the heads of two operators, climbed a wooden staircase, and went through the roof. The engine was reduced to junk.

> The above incident was an early warning of the importance of vibration and fatigue in propeller operation. At my insistence, further "routine" propeller tests on experimental propellers were made in a specially constructed "bombproof" shelter. All metal propellers of that time (1921-1922) failed, with murderous results to the engine. In one case the whole assembly of crankshaft, rods, and pistons was pulled out and thrown 20 feet from what remained of the engine and stand.

Three types of propellers emerged by 1923: fixed pitch, adjustable pitch, and reversible pitch. The materials used in making propellers included wood, metal, and micarta. With the development of large aircraft and airships powered by large engines, or combinations of engines, it was necessary to use propellers of very large diameter turning at low speeds. If such propellers were constructed of wood using

111

the conventional methods applicable to small propellers, they would be too heavy. Therefore, the division developed a method of constructing very large diameter wood propellers using a balsa wood core and then building up the necessary skin thickness of stronger woods by applying strips of very thin veneer. Propellers built by this method weighed less than half as much as propellers of conventional construction.[116]

Thus, a fixed-pitch propeller 17 1/2 feet in diameter was made from balsa wood for Goodyear's semi-rigid airship, the ZR-1 (the Navy's Shenandoah.) The wood base weighed some 39 pounds. With the covering, made up of 35 coats of one-fiftieth inch mahogany veneer, each of which was brushed with glue and ironed on, the total weight was only 110 pounds. In contrast, a propeller made from oak would have approached 450 pounds.[117]

Wooden propellers tended to warp, something neither micarta (molded composition) nor metal propellers did. Although until 1923 the Propeller Branch regarded metal propellers as still needing to establish their efficiency and practicality, a drop-forged aluminum blade had passed both whirling and flight tests. The aluminum propeller designed by the Propeller Branch was made of two standardized, interchangeable blades attached to a metal hub. The interchangeable blades permitted replacement of a broken blade without discarding the other one. The blades of the propeller, made in a set of forging dies, were drop forged to the finished size. The process produced the largest aluminum alloy drop forging known to the Propeller Branch. The metal was heat treated for maximum strength. The weight of the aluminum propeller was slightly more than wood, averaging about that of micarta. The Navy's use of metal propellers in the 1923 Pulitzer Trophy and Schneider Cup races added about 10 miles per hour to the speed of their aircraft, leading to service testing of the propellers. By 1925, drop-forged aluminum alloy propellers for Liberty engines had proven very satisfactory, increasing the speeds of all the aircraft in which these propellers were tested. These propellers proved very adaptable since it was possible to use the same design on aircraft ranging in speeds from 90 to 140 miles per hour. In large quantities, the aluminum propeller was potentially much less costly than wood.[118]

The Propeller Branch in 1923 considered the immediate field for reversible propellers was with dirigible balloons and for adjustable propellers with supercharged engines. The variation in angle on the adjustable propeller was from six to eight degrees. The angle on reversible propellers could change up to 45 degrees at which point it became a pusher instead of a tractor. The principle of operation in a reversible propeller was turning two separate blades in ferrules that formed part of the hub. The metal hub and ferrule contained the operating mechanism that could change the blade angles in flight. The pilot controlled the angle of the blade from the cockpit.[119]

Adjustable-pitch propellers enabled aircraft to take off with maximum climbing speed and then operate at optimum speed at altitude. Adjustable propellers assisted greatly in flying at high altitude with a supercharger since the supercharger kept the engine's power normal and the propeller produced more "bite" in the rarified air, almost reproducing flying conditions at sea-level. Without an adjustable propeller, the advantage of the supercharger would be lost. Reversing pitch was useful for small fields since the thrust of the propeller could be made to act in the opposite direction and function as a brake, reducing the stopping distance of the aircraft. By 1925, the division introduced adjustable pitch propellers with wood blades for engines of high horsepower. These propellers had been sufficiently refined that it was possible to build them for use up to 200 horsepower. The mechanism of the propellers allowed the pitch of the blades to be changed easily while flying.[120]

Although wood propellers were still dominant in 1926, metal ones were rapidly displacing them as standard on all service aircraft because of their greater efficiency. Tests at McCook Field and in England indicated that metal propellers gave better performance. Several factors were operating against using wood: tip speeds were increasing, greater power was being exerted on the blades, supercharged engines required variable pitch as did the demands for maximum performance in height and speed, the effects of weather and climate, and the scarcity of suitable wood. Metal propellers were made of steel or alloys of aluminum. Magnesium was less attractive because it was affected by moisture.[121]

In addition to the one-piece duralumin propellers of the Curtiss-Reed type, the detachable blade type by the Standard Steel Propeller Company of Pittsburgh, with blades adjustable on the ground, greatly simplified the problem of supply. Changing the pitch of blades in flight was another development successfully applied at McCook Field to a controllable-pitch propeller for the Liberty engine. The detachable blade principle, introduced by McCook Field, also was being applied to propellers with magnesium and micarta (bakelite) blades.[122]

Steel Propellers. Steel propellers had been used since 1917 in Europe. Steel had the advantages of being cheap, readily available, and easily worked. The blades were fitted into hubs and were capable of being adjusted in pitch or replaced if damaged. Blades were built up from sheared sheet steel pressed to shape and welded together at the edges. Riveting was not used. Welding was done parallel to the radius to avoid stressing the blades, and the weld at the root, where all the sheets entered the holding sleeve, was under compression. The sheets in the blade were of different lengths, giving a taper in thickness toward the tip. The laminated construction increased the strength of the finished blade over its component sheets, and assisted in damping out vibration. Each laminated sheet of the blade was continued into the attachment flange and the inner laminations were shaped so as to bring the center of gravity of the blade as close to the shaft as possible.[123]

This construction allowed making hubs that would take a wide range of blades. Hubs could be made to accommodate two, three, or four blades, and any of these propellers could be built up from stock. Similarly, balancing could be done with sufficient exactness that damaged blades could be interchanged with blades from stock without having to rebalance the propeller.[124]

Steel propellers were heavier than wooden ones, but the increased weight did not adversely affect the crankshaft, the end-bearing, or the performance. Steel propellers also gave consistent results in testing, something impossible to obtain with wooden ones. Steel propellers did not tear out at the hub or split at the weld. The Royal Air Force used steel propellers extensively. Early blades were in service for years without any trouble, and later blades were tested at 200-percent overload without failing. Samples cut up after two years of service showed no signs of fatigue in the steel or welding. Moreover, steel propellers resisted machine-gun fire better than wood. In general, the Lietner-Watts steel propeller demonstrated good performance, reliability, and durability.[125]

The largest propellers of steel were two-bladed propellers for the 700-horsepower Rolls-Royce Condor engine. They were 16 feet in diameter, rotated at 1,000 rpm, and had a peripheral speed of 840 feet-per-second. This type of construction could be used for tip speeds up to 960 feet-per-second. But for higher speeds, the Leitner-Watts propellers had blades of solid duralumin fitted into the standard steel hubs. Metal Propellers, Limited, in England, manufacturers of the Leitner-Watts propellers, was also developing a variable-pitch propeller in 1925.[126]

Reed's Duralumin Propeller. Dr. Sylvanus Albert Reed was engaged in high-frequency research in 1915 using an apparatus with a shaft turning at 36,000 rpm. Aware that problems were believed to occur when an object moving through the air approached the speed of sound, a belief based on experience with artillery shells, Reed decided to conduct experiments with high-speed propellers. Reed postulated that the difficulties encountered when the speed of sound was reached might depend on the shape of the moving object. He conducted a long series of experiments, beginning with propellers 20 inches long that he revolved at 14,000 rpm. Reed concluded that the supposed physical limit at the speed of sound did not exist and that the thrust of the propellers did not undergo appreciable variation when they exceeded that limit. Reed's theory was revolutionary at the time. As late as 1919, the British Advisory Committee for Aeronautics reported that its experiments showed that thrust practically disappeared at high tip speeds. McCook Field's testing corroborated the British findings. However, Reed succeeded in interesting Curtiss Aeroplane and Motor Company in his research, and the company placed facilities at his disposal.[127]

Reed found that duralumin propellers did not straighten out even if the propellers were made very thin, a feature he became aware of during his 1917 experiments. After testing small models in his laboratory, he completed work in 1920 at the Curtiss company with full sizes and power. Reed's experiments resulted in the Z-1 propeller of 1921, a propeller with very thin tips and razor-sharp edges. To achieve the thin sections, Reed's propeller was made of aluminum in one forging. It gave excellent results, producing very high efficiency at previously unheard-of rotation speeds. The first flight test was made in August 1921 with a 9-foot duralumin propeller on a Curtiss 160-horsepower aircraft. The pilot reported an improvement in speed compared to wooden propellers. In December 1921, McCook Field tested the same propeller during a successful 30-hour ground test at 50-percent overload; that propeller continued in flying service through 1924. Several other Reed propellers were also tested at McCook Field. Meanwhile, in 1922, Dr. Reed tested 3-foot models at Stanford University's wind tunnel, and reported the results in 1923. Reed's propeller became the preferred accompaniment to Curtiss' direct-drive D-12 engine.[128]

Duralumin propellers were made from a one-piece solid forging with the approximate shape of an untwisted propeller. The blank or flat slab was annealed and the section then shaped with milling cutters. The single piece of rolled and annealed duralumin plate was between 7/16-inch and 1/4-inch thick; tapered; camber faced and twisted in a rotary press for pitch, and then heat treated in a salt bath and quenched, and a shaft hole drilled. The propellers were finished by polishing and shaping of the edges. After the hub bosses were fitted to the center of the blades, the propellers were balanced and assembled. The result was a single-piece metal propeller without hollow spaces, welds, or rivets. At first, the duralumin propeller was made interchangeable with the same hubs used by the wooden propellers that it replaced, filler blocks being placed between the hub flanges and the propeller face. But using a special steel hub fitted to the duralumin propeller saved weight and was simpler. Although duralumin propellers were integral, they could be adjusted to some extent after trial, another advantage compared to wood. By 1926, the metal propellers were also being made in a detachable blade type, an adjustable-pitch type, and three- and four-blade types.[129]

Duralumin as used in the Curtiss-Reed propellers proved more suitable than wood. It could be used for both hollow and solid blades, and it was easily machined. The duralumin propeller exceeded all other propellers in simplicity, durability, ease of repair and restoration if bent, and economy. The principal advantages, however, were a significant gain in efficiency compared to wooden propellers and the capability of operating at tip speeds beyond the practical limits of wood propellers. Dr. Reed's study showed that above 900 feet per second tip speed, the efficiency of wooden propellers fell off rapidly. Moreover, wooden propellers at very high velocities were not safe for use whereas tip speeds of duralumin propellers were practical even with tip speeds exceeding the speed of sound. Varnish or an anodical process protected duralumin from atmospheric corrosion and sea water. Although duralumin was relatively expensive in 1926, the cost was expected to decline.[130]

Reed's patented duralumin propeller became known as the Curtiss-Reed propeller. The Curtiss company used them on racing aircraft where wooden propellers would have been impractical. In 1923, world speed records for both land and sea planes were won by Curtiss aircraft and engines with the duralumin propellers. Two CR-3 seaplane racers equipped with D-12 engines and the new propellers took first and second place in the Schneider Cup Race in England. A few days later, Navy R2 C-1 racers equipped with D-12A engines and Curtiss-Reed propellers won first and second places in the Pulitzer Trophy Race, the winner achieving 243 mph. Later, the Navy boosted the world speed record to 266.6 mph with the R2 C-1 aircraft. Dr. Reed also licensed foreign manufacturers. In December 1924, a French aircraft using the Reed propeller won the world speed record at 278 mph. After that, duralumin propellers were widely used.[131]

In 1924, the U.S. air mail service fitted Curtiss-Reed propellers on its DH-4 aircraft with Liberty engines. By the end of 1924, the Army had used these propellers on 15 Curtiss PW-8 pursuits for a year. The Navy had also used several Reed propellers continuously since early 1923. In the 1923 National Air

Races, only the Navy's two racers were equipped with duralumin propellers, but in 1925, 50 of 87 aircraft in the competition used Curtiss-Reed propellers. The Curtiss-Reed propellers was used extensively and continually gave better results than wooden propellers.[132]

Meanwhile, a single piece magnesium propeller was also produced by Dr. Reed. Magnesium alloy weighed only two-thirds as much as duralumin. First flight of an aircraft using the magnesium propeller occurred on 6 February 1925 at the Curtiss Company. The company's pilot flew a J-1 aircraft equipped with a Curtiss C-6 engine. Earlier, the workers tested the propeller on the McCook Field block. The tests established that magnesium alloy possessed the physical qualities needed for propeller construction. But ultimately magnesium did not prove as effective as aluminum.[133]

The duralumin propeller was recognized as the first important advance in propellers since standardization of the laminated wood propeller. In 1926, Dr. Reed won the Collier trophy, given each year for the most valuable contribution to aviation. In accepting the Collier trophy, Dr. Reed credited several individuals and organizations for assisting him. Among them was[134]

> *the U.S. Army Engineering Division of the Air Service, at McCook Field, of whom I will name [Major John F.] Curry, [Major Leslie] MacDill, [Frank W.] Caldwell, and [M.A.] Smith, who have made for me a long series of tests or experimental ideas, from which results, whether success or failure, I have greatly profited.*

On the eve of the move from McCook Field to Wright Field in 1927, <u>Slipstream</u> published a list of Ohio's aeronautical accomplishments. The list included many developments that benefited both military and commercial aviation. Among them were many contributions of the Power Plant Section and the Propeller Branch:[135]

- Refinement of the Liberty engine.
- Inversion of Liberty engine: high thrust line improved vision and maintenance; used on Loening amphibian.
- Geared Liberty with 2-to-1 reduction: enabled large single-engined aircraft to take off from small fields.
- Model W-1 18-cylinder engine of 750 horsepower.
- Large-bore air-cooled cylinder over 50 horsepower: basis of all air-cooled engine development in U.S.
- Salt-cooled valve: lengthened valve life more than 500 percent; decreased frequency of overhaul.
- Geared-type supercharger.
- Standardization of aircraft fuel systems and installation parts.
- Almen "barrel" engine: vibrationless engine of minimum frontal area.
- Air-cooled Liberty engine: first V-type air-cooled aircraft engine in U.S.
- 400-horsepower air-cooled radial engine: Curtiss R-1454.
- Four-engine Liberty transmission.
- Two-engine Liberty airship transmission.
- Side-type supercharger: 20,000-foot model.

116

- Detachable blade drop-forged aluminum alloy propeller: duralumin.
- Micarta propellers with fixed and detachable blades.
- Detachable hollow-blade steel propeller.
- Single-piece hollow-blade steel propeller.
- Aluminum leaf coating for wood propellers.
- Detachable power plant unit preceded foreign developments.
- 17 1/2-foot wooden propeller with balsa core.
- Adjustable- and reversible-pitch propellers with wood blades.
- Non-reflecting color for propellers used in night flying.
- Tapered-wing and tunnel-type radiator.
- First and only propeller testing laboratory of its kind in the U.S.
- Synchronizers, both mechanical and electrical, to permit a gun to fire through whirling propeller.
- Hand and electric starters for aircraft engines.
- Piston alloy and heat treatment for alloys of aluminum and magnesium.
- Heat-resistant enamel for aircraft engines.

The principal legacy of McCook Field was in establishing the importance of a strong, centralized aeronautical engineering organization to promote military aviation technology. The Engineering Division played a major role in transforming aviation into a serious engineering and scientific enterprise during the 1920s. The division became the most influential organization in American aviation during that decade, providing a start for some of the most talented men in the industry and setting standards for the future.[136]

Acquisition Policy Evolves

Although aeronautical experimental activities in WWI began too late to have an appreciable effect on combat operations during the war, McCook Field emerged as an effective center for research and development. For several months after the armistice, until the end of fiscal year 1919, the Air Service was able to continue wartime expenditures. During the 1920s, progress continued despite the reduction in funding. Money was invested chiefly in laboratory facilities and equipment at McCook and in a substantial inventory of experimental models of airplanes and equipment. Many experimental models were built before the armistice though they were not placed into production. Within three months after the armistice, the aircraft industry in the U.S. declined to 10 percent of its wartime strength, and was struggling to survive. The government emerged from the war without a procurement policy and it failed to provide money for new designs or new aircraft.[137]

Within a year after the armistice, a major national periodical was already vehemently protesting McCook Field's "excessive centralization," complaining that *With the American aviation industry at a standstill, twelve hundred men are at work at Dayton, many of them at very high salaries, and the plan of moving to Moraine City is constantly being brought up.*[138]

A reliable authority at McCook Field responded to this attack, attributing the situation to a lack of communication between the industry and military officials. Conceding that American designers were perfectly capable of designing excellent airplanes if encouraged, the source added:[139]

> *The Engineering Division feels that civilian designers have sacrificed ease of production and maintenance for performance. In general, it is true that experimental airplanes delivered to McCook Field are unsatisfactory for military purposes without extensive alterations.*

This anonymous source also defended the field's designing and building of aircraft, which he pointed out occurred only in a limited way. In that process, military experts learned how to assess aircraft submitted by outside designers and correct defects in these airplanes after they are placed in service. Even though McCook Field was willing and anxious to receive designs from civilian firms, a shortage of funds prevented the letting of many experimental contracts.[140]

In the years 1919, 1920, and 1921, Engineering Division reports to Washington dealt mostly with experimental development and testing rather than standardization and production. The division was designing and occasionally constructing experimental pursuit, attack, and observation aircraft; studying and designing other aircraft for night bombing, night attack, ground attack, and infantry liaison; working on air-cooled engines, cooling systems, and superchargers; and testing (and sometimes independently designing) of parachutes, leak-proof tanks, photographic equipment, radio, aerial torpedo, armament, and bombing equipment. Through 1922, McCook Field engineering designed 14 models and built a total of 27 aircraft from these designs, including bombers, attack, pursuit, observation, and a racer.[141]

The chief of the Engineering Division, Major Bane, regarded the process of awarding contracts for aircraft manufacture as needing improvement. The armistice ended cost-plus contracts for manufacturers, and led to the wholesale elimination of aircraft firms. In the early 1920s, one problem was painfully clear: originators of aircraft designs were denied proper compensation when the Air Service awarded production contracts based on their designs to competing firms. During his farewell remarks in 1922, Bane outlined areas needing reform but even he would not honor proprietary claims:[142]

> *Some day I hope to see a better system in the respect to the awarding of Government contracts to the various airplane manufacturers. The present system is not a good one, and it will be a great step toward aeronautical development when contracts are awarded to standard airplane manufacturers. At present, most of the manufacturers of aircraft who secure contracts with the Government, require financial aid from the Government before they are able to make production. Furthermore, the awarding of contracts is not confined to the original factory of the particular type of plane wanted, but may be taken over by any one of a number of manufacturing concerns who claim they are equipped to turn out the planes at a certain figure. For instance, some time ago bids were opened by the Government for a number of Martin Bomber planes, but the award did not go to the original Glenn Martin Factory, Cleveland, Ohio, but to another concern who claimed*

that they could manufacture Martin Bombers at a better figure. Now the right thing to do is first let the representatives of the different makes of planes bring their designs, etc., to the Engineering Division, and let these designs and other data be inspected and passed upon by the Engineering body. If a particular plane looks good, let the manufacturer send one to the Engineering Division station to be tested out. If it passes the test, then if practical, an order for the desired number of planes can be placed with the company. I do not think there can be the right kind of progress in aeronautical development until some plan like this is carried out, and what is more, aircraft manufacturers will never do much good until they become properly financed and commercialized.

On the other hand, I do not think that such a thing as "propriety" should enter into the question of contract awards. Contracts should be awarded solely on the merit of the product and upon the sound, sane judgment of the Engineering Division.

Postwar experimental activities between 1919 and 1923 were especially difficult in the face of severely reduced budgets, a growing indifference toward national preparedness, and increasing pressure for standardization of production. The Engineering Division's R&D work faced resistance from higher authority in the Army, even from within the Air Service. The enduring tension between logisticians and engineers over their classic approaches to research and development came to the fore early in McCook Field's history. In 1921, a memorandum from the Supply Group in Washington objected to the Engineering Division's emphasis on experimental equipment and called for standardization so that the Air Service would have equipment for an emergency, warning:[143]

If we continue this policy of buying a dab of every kind of experimental type of equipment that the Engineering Division in Dayton passes upon with the idea of conducting a service test, it would not be long before the entire Air Service would be engaged in service test work, and, should an emergency develop, it would be impracticable to put any kind of an organization in the field with the standard type of equipment.

Colonel Bane's correspondence with an English colonel in 1922 described the obstacles McCook Field needed to overcome in performing its R&D mission:[144]

I presume you are having the same difficulty that we are in these days of economy. Unfortunately the Government is not in the hands of scientists and when an economy wave strikes, experimental development is the first thing to suffer. We rather fear that the next year will be a very trying one in a financial way...I fear we shall have to discharge a great many engineers.

Some congressmen complained in 1922 that too many engines were being developed by the Air Service. Bane, as head of the Engineering Division, quickly countered that charge. His response cited 11 engines (both water- and air-cooled) in development, and argued that they were limiting their development to the bare necessities of the Service. These engines were needed, he said, to keep up with the rapid advance in design.[145]

Manufacturers were especially angry that the Engineering Division could give their aircraft designs to other firms for production. Companies that invested in design bureaus only to lose production contracts to the lowest bidder regarded this treatment as unjust. Manufacturers also denounced McCook Field's activities -- especially the design and construction of aircraft -- as infringing on their prerogatives. In the absence of a flourishing commercial sector, manufacturers needed military business to survive. Their protests against McCook Field's "foreign engineers" who were designing aircraft for the Air Service began immediately after the Great War. In 1920, manufacturers successfully blocked British dumping of WWI aircraft and engines on the U.S. market, but they could not stop the sale of the U.S. government's own surplus aviation equipment. Their protests against both McCook Field as an experimental factory and the Naval aircraft factory (authorized in July 1917 at the Philadelphia Navy Yard) finally succeeded. In response to mounting political pressure, General Patrick revised the Air Service's procurement policy in 1923. He explained his new policy for McCook Field in these words:[146]

> *I became convinced that the contentions of the manufacturers were valid and that it was essential to change radically the Engineering Division's practice. I decided that we would build no more airplanes at the Division, and, further, that no more aircraft designs would be created there. We would still maintain a designing staff, but its functions would be to pass upon the designs submitted to the Air service, while it would be available for consultation with outside designers, manufacturers, and those who had ideas to propose.*

McCook Field's reaction to criticism of its post-WWI development policy for aviation equipment was reflected in a detailed report, The Engineering Division: A Consideration of Its Status with Respect to the Air Service and the Aeronautical Industry in the United States and Abroad, published at the end of 1924. The division defended its activities in the immediate aftermath of WWI by noting that, at the time, neither industry nor the Air Service knew exactly what kind of military equipment was needed until more experience was gained. Admittedly, by the end of 1924, U.S. aviation had progressed beyond those limitations. The division attributed this progress in large measure to the fact that[147]

> *companies have retained, developed, and acquired (in many cases from the personnel of the Engineering Division) a corps of competent designers and contractors who are able to design and construct satisfactory aircraft and engines to meet Air Service requirements.*

In 1924, the mission of the Engineering Division in support of Army aviation reflected General Patrick's new policy. The new mission statement dropped the reference used in 1919 regarding design and development of aircraft and engines:[148]

> *The Engineering Division is specifically charged with the responsibility for the research and development work, the creation, development, and improvement of all aeronautical material and equipment used by the Air Service. It translates in addition, the requirements of the Service to the Aircraft Industry. The immediate objective of this work is the creation of the most efficient Air Service equipment with a view of having in readiness for immediate production and manufacture.*

By 1924, the Engineering Division ceased designing and building new aircraft and engines, and focused instead on monitoring the designs and production of manufacturers. Besides cooperating and coordinating with the various government branches, the division actively encouraged the aircraft industry. The division's revised approach applied not only to the manufacturing of aircraft, but also to aircraft engines and accessories. Virtually everything necessary for equipping military aircraft was turned over to industry to manufacture. As explained by Major John F. Curry, who became commander of the Engineering Division on July 1, 1924:[149]

> *...but having no desire to foster a spirit of competition with commercial industry this [new program] was now prosecuted in coordination with various manufacturing concerns. The Engineering Division thus became the clearing house between the Air Service and the manufacturer, interpreting for him the needs of the Army in specifications and drawings of articles to be built, testing the products when they were completed, refusing them if they did not come up to specifications; or if they did and were still unsatisfactory, pooling with the manufacturer engineering experience and suggestions in hopes of obtaining a better functioning product. Some remarkable developments have been the result of this cooperative effort.*

Thus, the division turned over all experimental data obtained at the Air Service's expense to the industry. One example cited by the division was the high-powered, air-cooled engine order from Curtiss Aeroplane and Motor Company, Incorporated. The Aeronautical Board determined that the Air Service should pursue development, but representatives of aircraft manufacturers in 1921 at a meeting of the Society of Automotive Engineers were reluctant to develop large air-cooled engines. The Engineering Division did the research and after several years turned the engineering data over to the manufacturer.[150]

People at McCook Field responded heatedly to accusations that they were overpaid and cramping the style of industrial firms that supposedly could provide the flying personnel with better aircraft, and more of them.[151]

Such allegations were especially objectionable when they came from foreigners like publisher Charles G. Grey of England. After he visited this country briefly in 1924, Grey observed that the lack of an aviation policy was hindering progress in the U.S. He added caustically: *"The Army seems to suffer from a place called McCook Field...McCook Field costs millions of money (dollars or pounds) and produces nothing."* Grey attacked government production as the *"least efficient, most expensive, and least progressive way of producing."* Those charges, however, were only an introduction to his principal indictment of McCook Field, its acquisition policy:[152]

> *But the chief trouble of all is the method of purchase. When at last a trade design is approved and somebody who has the necessary authority decides to order that design, instead of orders for that particular aeroplane being given to the firm which designed it and instead of agreeing with the firm a price which is high enough to cover experimental and designing costs, the order is put up to open tender.*

Thus, firms that invested money in new designs were left without orders and no income to produce fresh designs. Moreover, after an order was finished, the company's work ended and the trained workers were let go. Grey also condemned the Naval Aircraft Factory's role as equally hurtful toward manufacturers.[153]

After further reflection, Grey offered a more balanced assessment of McCook Field's role, conceding toward the end of his series of articles,[154]

> *There is no doubt that the work done by McCook Field, the Naval Aircraft Factory and Langley Field is all very necessary. Somebody apart from the [aviation] Trade ought to stand by and watch engines shimmy themselves to bits. Some Government official ought to contract myopia through squinting at wind-tunnel instruments. Some engineer or other ought to build an odd aircraft or so once in a while just to give the Government an idea of how much can be spent on such work -- it gives the Trade a chance of putting on high prices, for the high cost of Government work averages up against the low bids of prospectively bankrupt price-cutters. Therefore the work of those three establishments seems very necessary.*

On the other hand, Grey was adamant about the U.S. acquisition system needing reform along the lines of England's approved list of aircraft manufacturers.[155]

> *...American airplanes and air-engines are for the moment better than European aeroplanes and aero-engines and that American pilots and mechanics are at least as good as any on the other side of the Atlantic. But the fact remains that most American aircraft factories are without work and that the style of the American Flying Services is cramped by lack of aircraft and an insufficient number of officers and men.*

He considered outrageous the law that decreed *"all orders for aircraft must be given to the lowest bidder in open competition because no aviator should be made to fly the cheapest aircraft instead of the best. The present method is the surest way of killing American aviators first and killing manufacturers afterwards."* Attacking the Sherman Anti-Trust Law, Grey argued:[156]

One knows that the Army and Navy officials themselves are all in favor of forming a selected list of "approved firms" and of keeping those firms alive by judicious distribution of orders so that when the War comes each of those firms will be able to act as nucleus (or mother-firm) from which other factories may be taught how to build aircraft. But these officials cannot place orders as and how they will, because they are bound by the Law to order from the lowest bidder even though they may be morally sure that the lowness of the bid will break the firm and stop the fulfillment of the contract.

The result is that a firm may produce an excellent design and be paid for its experimental work and then have its plant flat for a year or more while a rival who has under-bid on its own designs is slowly going broke on an impossible contract.

Carrying this argument further, Grey cited the fact that the government honored the proprietary rights of producers of aircraft engines as the basis for treating aircraft manufacturers with the same equity.[157]

Even as it is the proprietary idea seems to exist in the aeronautic bureaux, for aero-engines are ordered from the makers without asking for bids. And if engines are proprietary articles why not aeroplanes? Any automobile plant could make D.12s, or T.3s, or anything else -- after a fashion. So why are they not asked to bid? The sooner the Services remove this anomaly by regarding aeroplanes as proprietary articles the better. Then the firms which design the good stuff will get the orders for it. And firms which make good stuff but are not good at design can have orders for the designs of other firms on a royalty basis.

Soon after General Patrick's revision of acquisition policy, the attacks on McCook Field ended. By the middle of 1925, the editor of <u>Aviation</u> withdrew his criticism that McCook Field was trying *"to meddle too much with aircraft production."* Instead, the periodical recognized that *"Lately there has been a great change and the development of aircraft has been turned over to the manufacturers."* The Army and Navy were praised for giving the U.S. the world lead in development of aircraft engines, but the Naval Aircraft Factory still came in for a bashing from the periodical: *"the aircraft industry might almost be termed a fabricating shop for naval aircraft."*[158]

At the end of 1925, <u>Aviation</u> acknowledged that the Engineering Division *"has recently been turned in the direction of making the dreams of the research workers come true."* The division was no longer *"trying to produce a freak device,"* but was trying *"to translate into service types of aircraft and accessories, the knowledge that has been gained in all parts of the world."* Admitting that the periodical had been quite critical of McCook Field in the past, the editor conceded:[159]

A careful study of the work being done at the headquarters of the Air Service Engineering Division, gives the impression that it is no longer attempting to concentrate within its borders a monopoly of aeronautical knowledge. Quite a contrary condition seems to prevail. Dissemination of information is the rule now. Every project also has, as its objective, some service purpose.

123

In testimony to the senate in February 1926, following investigations by the Lampert committee and the Morrow board, General Patrick pointed out that aviation firms lacked protection because they had no way to patent their aircraft designs. When designers brought proposals to the Air Service, General Patrick was obliged to advertise for bids to construct the aircraft, taking the private designs and putting them up for open bids. He protested that this process kept him from buying at a fair price in the way that other businessmen did.[160]

In nearly every single instance the designer has named a price higher than some of his competitors, because he knew more about the actual work than the others, and in practically every case where his competitors have taken the business away from him, they have lost money.

General Patrick also pointed out that his audit section audited every aircraft contract let by the Air Service, an assurance that no excess profit would be made. Thus, negotiated contracts with competent firms did not preclude competitive bidding. With passage of the Air Corps Act on July 2, 1926, the Army and Navy secretaries were permitted to exercise these contracting options; the act also provided for auditing of the contractors' books. Thereafter, the services were able to negotiate contracts for service testing a certain number of aircraft that were considered "experimental" since no design was fully developed until after passing a service test. The Army even negotiated production contracts with these manufacturers until November 1933, when, like the Navy, it was forced to let all production contracts through formal bidding.[161]

Lack of enabling legislation delayed the start of the five-year expansion program provided for in the Air Corps Act until July 1, 1927, causing the end date to be set back to June 30, 1932. The act authorized the Air Corps 1,650 officers and 15,000 enlisted, and set a goal of 1,800 serviceable aircraft, including the annual procurement of 400 aircraft. In the end, adequate funding was not appropriated because the Great Depression intervened and the goals were not realized. By June 30, 1932, the Air Corps had 1,305 officers and 13,400 enlisted with 1,709 aircraft.[162]

Meanwhile, under the Air Corps Act, three major divisions were established: Materiel, Operations, and Training. The Air Corps established the Materiel Division at Dayton, Ohio, on October 15, 1926. Brigadier General William E. Gillmore, previously chief of the Air Service's Supply Division in Washington, D.C., became chief of the Materiel Division. The new division incorporated functions previously performed by the Supply Division, the Engineering Division, the Industrial War Plans Division, and the Materiel Disposal Section of the Air Service. The operations of the Materiel Division were conducted by six major sections: Engineering, Procurement, Administration, Field Service, Industrial War Plans, Repair and Maintenance. Thus, the new division embodied the functions of engineering, procurement, supply, maintenance, and industrial preparedness.[163]

When the Materiel Division moved to Dayton in the fall of 1926, all materiel activities of the Air Corps were centralized in Dayton with temporary headquarters at McCook Field pending completion of Wright Field. General Gillmore had offices at McCook Field until Wright Field's more modern and spacious facilities were ready for occupancy in 1927.[164]

The reorganization of October 1926 subordinated the Engineering Division within the Materiel Division as the Experimental Engineering Section. The engineering section included seven main branches: Aircraft, Power Plant, Equipment, Materials, Armament, Engineering Procurement, and Shops. The mission of this section was to initiate experimentation, design, testing, and development of aircraft, engines, propellers, accessories, and associated ground equipment. From the dedication of Wright Field in 1927 until 1939, when the Materiel Division was restructured as the U.S. prepared for World War II, the field's engineering activities were conducted by the Experimental Engineering Section.[165]

Engineering Functions Relocated from McCook Field to Wright Field

During the Great War, McCook Field was established as a temporary emergency site for Air Service experimental work to meet wartime conditions. Langley Field, Virginia, was being constructed as a permanent installation. After the war ended, engineering activities were expected to relocate to Langley. By 1919, some permanent buildings (laboratories and hangars) were already built at Langley. On the other hand, Dayton had several clear advantages, including skilled labor, inexpensive housing, a central location, proximity to industry -- and the city already was the location of the Engineering Division. McCook Field's small size, however, meant that the division needed a larger site. But the War Department's requests to congress for funds and legislation to establish permanent engineering facilities went unanswered.[166]

In March 1919, the Assistant Secretary of War let it be known that the Army intended to abandon McCook Field, generating rumors that the Engineering Division would be relocated to a permanent site, presumably Langley Field. But as the year wore on, many other places bobbed up in the rumors that circulated, including the Dayton-Wright Airplane Company's facilities in Moraine, San Diego, Mineola in New York, and even McCook Field. Colonel Bane recognized that the situation at McCook was not ideal, especially because the growth of aviation made the facilities inadequate. Because most of McCook's facilities were temporary, they were fire hazards and costly to maintain. The field lacked a rail line, requiring supplies and equipment to be trucked in. The field was too small to accommodate the larger postwar aircraft under development. Aircraft tested at McCook also were a threat to the people living around the field. Furthermore, bombing tests must have proved particularly unnerving for nearby residents. Finally, rental charges for leasing the property increased each year.[167]

Bane's main concern was to have a decision made so that the division could focus on its work. He realized, however, that moving the division would cause the loss of experienced people and disrupt operations. As it turned out, consideration of Langley soon ended, and experimental activities were transferred from Virginia to McCook Field in May 1919. Other sites were considered but all lacked adequate facilities. The Dayton-Wright Airplane Company at South Field in Dayton offered an alternative site at a cost of several million dollars, but congress did not make funds available. In the absence of a decision at the end of 1920, Major Bane explained that plans were underway to keep the Engineering Division at McCook Field at least until July 1922. In 1921, Bane reported that a layout plan applicable to any reasonable piece of ground had been developed. He added that the economy practiced by the Federal Government prevented further discussion about a permanent site for the Engineering Division.[168]

In 1922, civic leaders in Dayton, reacted to the possibility of losing the Engineering Division. John H. Patterson, the founder of the National Cash Register Corporation, and Brigadier General Billy Mitchell, assistant chief of the Air Service at the time, discussed keeping McCook Field's operations in Dayton. Patterson, who supported Mitchell's call for an independent air force, lobbied for increased congressional appropriations for the Air Service and for keeping the Engineering Division in Dayton. When congress failed to appropriate funds to relocate McCook's operations, Patterson asked the Dayton Chamber of Commerce to host a luncheon on May 5, 1922 to honor Mitchell and Bane. At that luncheon, Patterson presented a plan for retaining McCook Field's functions in Dayton.[169]

Two days later, Patterson died. But his son, Frederick Beck Patterson, succeeded as head of NCR and took up the cause of keeping the Engineering Division in Dayton. Patterson formed a group of community leaders into the Dayton Air Service Committee by October. Other cities had already offered land to the government, and the committee quickly followed suit. The committee identified a site, largely owned by the Miami Conservancy District, adjacent to Wilbur Wright Field. The Air Service agreed to accept the land. At a dinner meeting on October 25, the Dayton Air Service Committee announced a public campaign to establish a new aviation facility as a monument to the Wright brothers. On October 31 and November 1, 1922, the committee collected over $400,000 in pledges. The committee immediately acted to transfer the land titles to the government. Even so, competing interests in congress caused delays in voting funds to relocate the Engineering Division.[170]

On the occasion of his retirement in December 1922, Major Bane expressed his support for the actions of the Dayton Air Service Committee:[171]

> In regard to the new location of McCook Field, I feel that there are no advantages great enough to warrant moving McCook Field and its personnel to a distant point such as some locations under consideration. Although the Dayton site may not be without its faults as a model location, it is as good as any of the other locations considered, and there is the added advantage of a short move. A good number of employees of McCook Field live here in Dayton; they own property here, and the necessity of selling out and moving with the Division would practically mean a disorganization. The Engineering Division is here; the land has been donated. I would say, stay here; and I believe the Government will decide to do so.

However, even after the committee purchased the land, however, opposition to keep the Engineering Division in Dayton, Ohio, continued. Political and military officials in Washington focused on relieving the anxiety felt by supporters in Dayton.[172]

General Patrick, the chief of the Air Service, emphasized in his annual report for 1923 that the rent for McCook Field was increasing rapidly and that city growth was making flying very hazardous as aircraft became larger and faster. Patrick recommended that the government immediately accept the land located several miles east of Dayton that the citizens' committee in Dayton had to offer. He also requested an appropriation of $1 million for preliminary improvements.[173]

In August 1924, the committee transferred ownership of 4,520 acres of land in Montgomery and Greene counties to the federal government. President Coolidge's acceptance of the land eased concerns in Dayton about the possibility of losing the Engineering Division. In 1925, congress began appropriating money for the construction of facilities and for purchasing equipment. In August 1925, the land donated to the government, including Wilbur Wright Field, was renamed Wright Field to honor both brothers. Groundbreaking ceremonies were held at Wright Field in April 1926. Thereafter, construction progressed rapidly. Elements began moving from McCook Field in March 1927. The move required 69 buildings to be emptied and demolished. By the time Orville Wright raised the flag at the dedication of Wright Field on October 12, 1927, the government had invested $1.5 million for the buildings and experimental laboratories. Another $2 million was required to complete the original plans for the new field.[174]

On the eve of Wright Field's dedication, General Patrick explained the mission of the Materiel Division at some length. Actually, his description captured the essence of the Engineering Division's activities at McCook Field during the previous decade and the tradition that engineers would follow at the new field:[175]

> *Putting it briefly, the Materiel Division investigates every piece of equipment which goes to make up the whole structure of the airplane and, through a series of exhaustive tests, determines the safety factor of each. Every possible precaution is taken to insure the safety of flying personnel. Nothing is left to chance; there is no guess work; everything done is based on careful calculations and thorough test, not only by men in the shops but also by expert test pilots who take the airplanes in the air and subject them to various maneuvers to satisfy themselves that they are safe for flying. This policy, in addition to protecting the lives of the pilots, prevents exorbitant expenditures by the Government for large quantities of airplanes which may be found subsequently unfit for use. No new type of airplane built for the Army Air Corps receives the Materiel Division's approval until the experimental model fulfills all the requirements of that division. It is now possible to determine with a considerable degree of accuracy, from data assembled since the establishment of this Division, the performance of any airplane even before it is built. Complete sets of specifications, drawings, and parts lists covering the latest design of each type of airplane, engine, instruments and article of equipment, are here available.*

CHAPTER 3 REFERENCES

McCook in the 1920s is discussed by Terence M. Dean, The History of McCook Field, Dayton, Ohio, 1917-1927 (Dayton, OH: University of Dayton, July 28, 1969), a master's thesis that provides a good starting point though inexact in places. An indispensable source is Slipstream, McCook Field's quasi-official periodical published magazine indispensable for understanding the Engineering Division and its times; it was published monthly (usually) between August 1919 and June 1928 -- but not at all in 1922 -- until it was bought out by Airway Age.

Useful publications by historians at Wright-Patterson Air Force Base include Bernard J. Termena, Layne B. Peiffer, and H.P. Carlin, Logistics: An Illustrated History of AFLC and Its Antecedents, 1921-1981 (WPAFB, OH: AFLC History Office, c.1982), and the illustrated volume by Walker-Wickam cited in Chapter Two. For propeller developments, see the oral history interview of Daniel Adam Dickey published in 1986 by Lois E. Walker, a historian at Wright-Patterson Air Force Base. Histories cited in Chapter Two by Claussen, Purtee, Frey, and Mooney and Layman, also apply to this chapter's discussion of the Engineering Division at McCook Field.

Three Air Force historians discuss the Engineering Division: Maurer Maurer, "McCook Field, 1917-1927," Ohio Historical Quarterly, LXVII (Jan. 1958), 21-33; Charles G. Worman, "McCook Field: A Decade of Progress -- Flight Test Center of the 20s," Aerospace Historian (Spring 1970), 12-15, 34-36; and Charles Joseph Gross, "McCook Field: Bellwether of American Military Aviation, 1917-1927," (draft article dated November 1990.) Summaries of McCook Field accomplishments are also provided by Maurer, "McCook Field, 1917-1927," Ohio Historical Quarterly, LXVII (Jan. 1958), 21-33; and Charles G. Worman, "McCook Field: A Decade of Progress," Aerospace Historian, (Spring 1970), 12-15, 34-36.

Two illustrated articles discussing McCook Field's contributions were published by Walter J. Boyne: the first part was "The Treasure Trove of McCook Field," Airpower, V:4 (July 1975), 6-25; the second part, oddly, was printed in a different periodical as, "The Treasures of McCook Field," Wings, V:4 (Aug. 1975), 8-25. Boyne pays tribute to the Engineering Division's pioneers, focusing especially on the aircraft developed. McCook Field published a popularized account of the Engineering Division's activities as they had progressed by 1924 in a 32-page brochure under the title A Little Journey to the Home of the Engineering Division, Army Air Service, n.d. [c. 1924]. Numerous photographs illustrate the field's activities and technical accomplishments as government support for the division was declining.

Several contemporary reports place McCook Field into historical perspective. An overview was provided in three parts by Gardner W. Carr, "Organization and Activities of the Engineering Division of the Army Air Service," U.S. Air Service, VI:6, VII: 1 & 2 (Jan, Feb, March 1922.) Another landmark report, this one for the mid-1920s, was issued by the Engineering Division: The Engineering Division: A Consideration of Its Status with Respect to the Air Service and the Aeronautical Industry in the United States and Abroad (Misc. Nr. 238), Eng. Div, AS, 20 Nov. 1924; this 85-page report is the best single source for understanding McCook Field's organization and functions after acquisition poicy was redirected by General Patrick, and provides details on employment and funding. At the end of McCook Field's existence, Brigadier General William E. Gillmore, chief of the Materiel Division, published a "Review of Year's Developments in the Army Air Corps," Slipstream, VIII:3 (March 1927.)

Especially useful are the annual reports issued during the 1920s, and often published by contemporary periodicals, by the chief of the Air Service and, later, the Air Corps, as well as by the chief and assistant chief of the Engineering Division. Annual reviews of aviation progress with insights about technical and commercial issues from the viewpoint of private industry, starting in 1919, are in the Aircraft

Year Book (New York: Manufacturers Aircraft Association, Inc., 1919 on); these reviews include many illustrations and considerable information on the military services, especially in the 1920s. One fundamental article for acquisition history is "Army and Navy Procurement," in the 1934 book. Who's Who in American Aeronautics (New York City: The Gardner Publishing Co., Inc.), published several times during the 1920s, is useful for key persons but not complete and not always reliable.

Contemporary insight into military aviation progress in the 1920s is provided by a number of other periodicals. Among these periodicals are U.S. Air Service (later Services), Air Service News Letter (later Air Corps News Letter), Aero Digest, and Aviation and Aircraft Journal (subsequently, just Aviation.) Specific articles from technical journals by Hallett, Heron, Jones, and others are cited in the reference notes below.

Discussion of engine technology the interwar period is provided by Robert Schlaifer, Development of Aircraft Engines (Andover, Massachusetts: The Andover Press, Ltd., 1950); the book is an insightful look that concludes U.S. private firms motivated by profits were better at details, development, and production -- notably after 1926 -- but that the military and civil service systems made the government superior at general goals and applied research. Included in the book is Samuel D. Heron's Development of Aviation Fuels. Dr. Sanford Alexander Moss, Superchargers for Aviation (New York: National Aeronautics Council, Inc., 1942), discusses the history and technology of the supercharger as it developed before World War II.

Technical background on engines was provided in several books. Charles Fayette Taylor, Aircraft Propulsion: A Review of the Evolution of Aircraft Piston Engines (Washington, DC: Smithsonian Institution Press, 1971), provides a readable primer on piston engine development. Taylor's book includes effective charts, illustrations, and bibliography. Another excellent discussion -- one lacking both bibliography and index -- is provided by Hugo T. Byttebier, The Curtiss D-12 Aero Engine (Washington, DC: Smithsonian Institution Press, 1972.) A personal history is provided by S.D. Heron, History of the Aircraft Piston Engine (Detroit, Michigan: Ethyl Corp., 1961); Heron, a participant in many developments related to aviation engines, discusses the piston engine and its related equipment including contributions by European technology. Even more instructive is Heron's Autobiography, edited by Robert V. Kerley from tapes (c. 1963), but unfortunately not published; Kerley provided an excellent bibliographical section.

Philosophical discussion of technology can be found in Ronald E. Miller and David Sawers, The Technical Development of Modern Aviation (London: Routledge & Kegan Paul, 1968), with emphasis on the technology of commercial aviation. Similarly, Joseph Lawrence Nayler, Aviation: Its Technical Development (Philadelphia: Dufour Editions, 1965), emphasizes developments in European propulsion, especially England's. For another aspect of the 1920s, emphasizing aviation in general, see Richard P. Hallion, Legacy of Flight: The Guggenhein Contribution to American Aviation (Seattle: University of Washington Press, 1977.

Military leaders provided insight into contemporary aviation progress in the 1920s from their personal experiences in several autobiograpies. Among them are books by Major General Mason M. Patrick, The United States in the Air (New York: Doubleday, Doran and Co., Inc., 1928); Major General Benjamin D. Foulois, From the Wright Brothers to the Astronauts (New York: McGraw-Hill Book Co., 1968); and General Henry H. Arnold, Global Mission (New York: Harper and Brothers, Publishers, 1949.) Patrick tends to be overly general, Foulois blasts Mitchell, and Arnold never forgot a good meal, but all lived through the 1920s.

For background discussion of the reorganization legislation in 1920, see David Allan Tretler, Opportunity Missed: Congressional Reorganization of the Army Air Service, 1917-1920 (Houston, TX: Rice University, May 1978), a thesis submitted for the degree of master of arts, passim. Tretler judged the 1920 law as precedent for the peacetime air arm, given the power of military tradition.

Two books by Grover Loening, Our Wings Grow Faster (New York: Doubleday, Doran & Co., Inc., 1935), and Takeoff Into Greatness (New York: G.P. Putnam's Sons, 1968), give useful insight from a contemporary manufacturer. Eugene Edward Wilson's two books provide a naval viewpoint, and tend to play down McCook Field's contributions, usually by omission. See his Air Power for Peace (New York: McGraw-Hill Book Co., Inc., 1945), and Slipstream: The Autobiograph of an Air Craftsman (New York: McGraw-Hill Book Co., Inc., 1950.)

[1]

Walt Boyne, "The Treasure Trove of McCook Field," Part I, Airpower, V:4 (July 1975), 12-13, 15. Walt Boyne, "The Treasure Trove of McCook Field," Part II, Wings, V:4 (Aug. 1975), 13, cited after this as Boyne by periodical. S.D. Heron, History of the Aircraft Piston Engine (Detroit, Michigan: Ethyl Corp., 1961), 118.

To assist with engine design, McCook Field had an engine museum by 1920 that was considered the most complete in the U.S. if not the world. The museum room, described as spacious and well lighted, had virtually every aviation engine of importance, including several types of Liberties, Sunbeams, Gnomes, and Mercedes. The engine display saved designers considerable research time, and McCook Field was urged to provide other rooms for propellers, wings, and aeronautical accessories. Leon Cammen, "McCook Field and American Aeronautics," Mechanical Engineering: The Journal of the American Society of Mechanical Engineers (Aug. 1920), 444, cited after this as Cammen article.

By 1923, the aviation museum at McCook Field had four large steel and concrete hangars with 26,000 square feet of floor space; one of the buildings was devoted exclusively to engines. The interior of the engine building was finished especially for the collection. Since the museum was intended primarily for designers and engineers, the engines were given a special display. Thus, cabinets behind each engine displayed completely disassembled duplicates for most of the engines. The disassembled engines -- thoroughly cleaned and polished -- were arranged in a pattern so that their respective parts were in the same relative position for each engine. A cylinder from most of the disassembled engines was displayed in sectional form with valves, pistons, push rod, and connecting rod. Often, special tools had to be used in their disassembly. The exhibit included 65 different types of engines (67 types by 1925), including models of almost all European engines used in World War I and models of dirigible engines. The engine types ranged from 4 to 16 cylinders, air- and water-cooled, stationary and rotary. Included were American, Austrian, British, French, German, and Italian engines.

In addition, the buildings housed 56 aircraft, squeezing them in by placing their detached wings behind the fuselages. Among these aircraft were 28 American, 14 German, and various other foreign models. Many still had the marks of WWI combat. T.C. McMahon was chief overseer of the museum. Boswell H. Ward, "The Air Service Technical Museum," Slipstream, IV:9 (Sep. 1923), 17-19. "The Aeronautical Museum at McCook Field," Air Service News Letter, IX:13 (July 28, 1925), 13-14. "The Aeronautical Museum at McCook Field," Aero Digest, VII:2 (Aug. 1925), 436.

[2]

Maj. Gen. Charles T. Menoher, Ch, AS, "Airplane Most Potent Weapon Ever Produced by Man," U.S. Air Service, V:3 (April 1921), 15. "The Reorganization of the Air Service," DMA Weekly News Letter, I:n.nr. (March 15, 1919), 1-2. "Air Service Reorganization," DMA Weekly News Letter, I:n.nr. (March 22, 1919), 1. "Statement from the Office of the Director of Air Service," DMA Weekly News Letter, I:n.nr. (March 29, 1919), 1-2. Purtee, 122, 125, 141.

[3]

Naval leaders, like Army leaders, opposed separating aviation from their control. From the onset of the argument over an independent air force, conservative military leaders preached cooperation rather than unification. See, for example, Franklin D. Roosevelt, Asst. Secy/Navy, "Why Naval Aviation Won," U.S. Air Service, I:6 (July 1919), 7-9.

Mitchell's public challenge that naval ships were vulnerable to aerial bombing led to sinking of four WWI German ships, a submarine, a destroyer, a cruiser, and the battleship Ostfriesland, in June and July 1921 off the Virginia Capes. The battleship was sunk by eight Martin MB-5 bombers with 2,000-pound bombs. In September 1921, the U.S. Navy's battleship Alabama was also sunk in day and night bombings. These sinkings actually helped the Navy's Bureau of Aeronautics, established in August 1921 under Rear Admiral William Adger Moffett, with its program for aircraft carriers -- and air-cooled engines -- and provided another major customer for the aircraft industry. Moffett, the equal of Mitchell in publicizing aviation, opposed a separate air force. Ironically, Moffett was in the position of having to sell the Navy on aircraft carriers. Fortunately, rivalry between Army and Navy pilots contributed to the growing excellence of U.S. aviation, including R&D. Disputes between the services over what the ship sinkings actually proved effectively kept military aviation in the public consciousness long after the bombing tests. The public tended to believe its eyes.

"Ostfriesland Sunk by 2000-lb. Aircraft Bombs," Aviation, XI:5 (1 Aug. 1921), 128-132. "Bombing Tests of the U.S.S. Alabama," Aviation, XI:14 (Oct. 3, 1921), 396-397. Brig. Gen. William Mitchell, "Talk Delivered at McCook Field, October 26, 1921," Slipstream, III:21 (Christmas 1921), 5-11. Capt. A.W. Johnson, USN, "Lessons from the Bombing -- A Navy View," U.S. Air Service, VI:3 (Oct. 1921), 29-33. "Navy Opposes Single Air Force," Aviation, XIII:8 (Aug. 28, 1922), 252. Terence M. Dean, The History of McCook Field, Dayton, Ohio, 1917-1927 (Dayton, OH: University of Dayton, July 1969), 92-96. Grover Loening, Takeoff Into Greatness (New York: G.P. Putnam's Sons, 1968), 149-153, 159. Eugene E. Wilson, Air Power for Peace (New York: McGraw-Hill Book Co., Inc., 1945), 91-93. Eugene E. Wilson, Slipstream: The Autobiography of an Air Craftsman (New York: McGraw-Hill Book Co., Inc., 1950), 1, 6, 8-11, 19-21, 23-27, 46, 58-60, 62-72. See also, "Admiral Moffett Claims All Sea Flying for Navy," Aviation, XVI:21 (May 26, 1924), 558-560.

Meanwhile, the British in February 1921 carried out bombing tests by aircraft on the German battleship Baden, sinking it at Spithead. Ironically, General Mitchell's call for using three German liners, the Leviathan, Von Steuben, and Agamemnon, as aircraft carriers was not funded. Speaking to the Society of Automotive Engineers in Washington, DC, in April 1921, Mitchell expressed himself more conservatively than his reputation suggested:

> In conclusion General Mitchell stated, in reply to a question, that the battleship was not yet obsolete, but that it might be superseded by aircraft within the next 20 or 30 years.

"Aero Club News," U.S. Air Service, V:3 (April 1921), 28. See also, "General Mitchell Tells House Committee Airplanes Can Destroy Battleships," U.S. Air Service, V:1 (Feb. 1921), 22.

On September 5, 1923, Mitchell's forces sank two obsolete U.S. battleships, the Virginia and the New Jersey, off Cape Hatteras, North Carolina. In the 1923 operation, bombs up to 2,000 pounds were used from altitudes of 10,000 and 11,000 feet by Martin bombers powered by two Liberty 12 engines. Effective support for these operations was provided by the Engineering Division. "Bombs Sink Battleships Virginia and New Jersey," Aviation, XV:12 (Sep. 17, 1923), 330-334. Maj. Gen. Mason M. Patrick, Ch, AS, "Without Adequate Air Force We Invite A National Disaster," U.S. Air Service, VIII:10 (Oct. 1923), 11-14. J.R. Moser, "Recent Aerial Bombing Tests on Battleships," Slipstream, IV:10 (Oct. 1923), 12-15. "Bombing Battleships from Airplanes," Slipstream, V:10 (Oct. 2-4, 1924), 69. Mitchell's speech at McCook Field in 1921, as printed in the Christmas 1921 issue, was reprinted substantially unchanged as controversy continued to swirl around him; see Brig. Gen. William Mitchell, "The Virginia Capes Bombing Test," Slipstream, VI:2 (Feb. 1925), 26-29. See also, Brig. Gen. William Mitchell, "Neither Armies Nor Navies Can Exist Unless the Air Is Controlled Over Them," U.S. Air Services, X:5 (May 1925), 15-18.

Mitchell's continued assaults resulted in his "exile" to San Antonio in April 1925 and replacement by Brigadier General James Edmond Fechet as assistant chief. Even in exile, Mitchell lashed out. Following the crash in Ohio on September 3, 1925 of the Navy dirigible Shenandoah (built in the Naval Aircraft Factory starting in 1920, it first flew on September 4, 1923), Mitchell charged defense leaders with "incompetency, criminal negligence and almost treasonable administration." Admiral Moffett responded to Mitchell, hinting at demagoguery and stating: "The most charitable way to regard these charges is that their author is of unsound mind and is suffering from delusions of grandeur." On October 28, 1925, Mitchell was tried before a court-martial under charges of violating the ninth Article of War; he was found guilty and sentenced to be suspended from rank, command, and duty for five years. He resigned effective February 1, 1926 to continue his efforts in civilian life. Mitchell died in February 1936.

Mooney & Layman, 68. Goldberg, 32. "Uncle Sam's Big Aerial Sisters: The Shenandoah and ZR-3," Slipstream, V:10 (Oct. 2-4. 1924), 57. "The General Mitchell Controversy," Slipstream, VI:2 (Feb. 1925), 5. "General Mitchell's Parting Address," Air Service News Letter, IX:8 (May 5, 1925), 3-6. "Admiral Moffett Replies to Accusations," Aviation, XIX:12 (Sep. 21, 1925), 353. "President Coolidge Names Board of Nine to Decide Aviation's Needs," U.S. Air Services, X:10 (Oct. 1925), 23-25. H.F. Ranney, "The Aviation Inquiry," U.S. Air Services, X:11 (Nov. 1925), 16-20. H.F. Ranney, "Colonel William Mitchell Declared Guilty," U.S. Air Services, XI:1 (Jan. 1926), 29-31. "The Mitchell Court Martial," Slipstream, VII:1 (Jan. 1926), 6. Loening, Takeoff, 161-162, 170. Wilson, Autobiography, 56-60, 70.

Meanwhile, Navy aviators like Rear Admiral William S. Sims advocated construction of carriers before the U.S. entered WWI, something the British had done as early as 1915. During the war, submarine warfare diverted U.S. naval attention from acquiring carriers. In 1919, congress appropriated money for converting the collier Jupiter to an experimental carrier. This first U.S. carrier, renamed the Langley, was completed in 1921. Based on experience gained with the Langley, two modern aircraft carriers followed. The new aircraft carriers (launched in April and October 1925, respectively) joined the fleet in 1927, the Saratoga on November 16 and the Lexington on December 14, converted from battle cruisers following the Washington Arms Limitation Conference. Under the leadership of Admiral Moffett, the Navy moved to overcome England's lead by building the largest, fastest, most powerful aircraft carriers. Meanwhile, the bureau was developing aircraft suitable for carrier use, emphasizing air-cooled engines to save weight and simplify maintenance.

Cmdr. Kenneth Whiting, USN, "The Langley -- A Floating Airdrome," U.S. Air Service, IV:6 (Jan. 1921), 8-10. Aircraft Year Book 1927, 78. Aircraft Year Book 1928, 132-136. "Highest Powered Naval Vessel in the World," U.S. Air Services, XII:11 (Nov. 1927), 23. "The Airplane Carrier 'Saratoga,'" Aero Digest, XI:6 (Dec. 1927), 715. "Our Naval Aircraft Carriers," Airway Age, IX:9 (Sep. 1928), 60-61. Wilson, Air Power, 92-100. Grover Loening, Our Wings Grow Faster (Garden City, New York: Doubleday, Doran & Co., Inc., 1935), 189-192. Wilson, Autobiography), x-xi, 4-5, 21-22, 78, 111-112, 114, 134-141, 146-148. Robert Schlaifer, Development of Aircraft Engines (Andover, Massachusetts: The Andover Press, Ltd., 1950), 167.

[4]   Purtee, 128.

President Warren G. Harding, elected in November 1920, committed his administration "to a period of economy and efficiency in government." To that end, congress provided him a Bureau of the Budget. The budget director, Charles G. Dawes called on federal employees to support improved methods, elimination of waste, and reduced expenditures.

[5]   Purtee, 112.

[6]   For background discussion of the reorganization legislation in 1920, see David Allan Tretler, Opportunity Missed: Congressional Reorganization of the Army Air Service, 1917-1920 (Houston, TX: Rice University, May 1978), a thesis submitted for the degree of master of arts, passim. Tretler judged the 1920 law as precedent for the peacetime air arm, given the power of military tradition. Maj. Gen. Mason M. Patrick, Ch, AS, "The Army Air Service," Aviation, XIII:23 (Dec. 4, 1922), 740-742; and XIII:24 (Dec.11, 1922), 777-779. Purtee, 124, 127, 135. Mooney & Layman, 47-48. Goldberg, 29, 36. Charles Joseph Gross, "McCook Field: Bellwether of American Military Aviation, 1917-1927," draft paper, Nov. 1990, cited after this as Gross paper.

[7] General Patrick, who called for government subsidies and regulation of commercial aviation, also recognized the military potential of aviation and publicized his views. Patrick supported an independent air force, suggesting establishment of a Department of Defense with a secretary at its head and sub-secretaries for land, sea, and air forces. Recognizing that achieving an independent air force would take time, Patrick came to recommend an Air Corps, comparable to the Navy's Marine Corps, as a necessary step toward that goal.

Maj. Gen. Mason M. Patrick, Ch, AS, "The World From Above," Aero Digest, VI:1 (Jan. 1925), 14-16, 47. "General Patrick Favors a Department of Defense," Air Service News Letter, IX:5 (March 17, 1925), 14. "General Patrick on Independent Air Force," Aviation, XVIII:24 (June 15, 1925), 662-663. "General Patrick's Statement Before Air Board," Air Service News Letter, IX:17 (Oct. 2, 1925), 1-2. "Extracts from the Annual Report of the Chief of the Air Service (Major General M.M. Patrick) to the Secretary of War, Fiscal Year Ending June 30, 1925," Air Service News Letter, IX:20 (8 Dec. 1925), 1-5; this data also appeared in Aviation, XIX:23 (Dec. 7, 1925), 804-805. See General Patrick's testimony to the House Committee on Military Affairs, "Would Have Air Corps Similar to Marine Corps," U.S. Air Services, XI:3, 4, & 5 (March, April, & May 1926), passim. See also, Maj. Gen. Mason M. Patrick, Ch, AS, "The Use of Aircraft in Future Wars," Slipstream, VII:5 (May 1926), 24-26. Patrick's autobiography, passim.

Major General Fechet, who became assistant chief under Patrick on 27 April 1925, succeeded Patrick as chief of the Air Corps. Colonel Benjamin D. Foulois succeeded Fechet as assistant chief of the Air Corps, becoming a brigadier general. "New Assistant Chief of Air Service," Air Service News Letter, IX:5 (March 17, 1925), 7-8. Frederick R. Neely, "General Fechet to Succeed General Patrick," U.S. Air Services, XII:7 (July 1927), 34-36. "General Patrick Retires," Aviation, XXIII:24 (Dec. 12, 1927), 1402-1403. Sgt. O'Connor, "Brig. Gen. Benjamin D. Foulois," Aviation, XXIII:25 (Dec. 19, 1927), 1466. Aircraft Year Book 1928, 123, 125-126, 327.

Contemporary assessment of General Patrick's six years as chief of the Air Service and the Air Corps was provided by Frederick R. Neely, "The Air Corps Under General Patrick," U.S. Air Services, XII:6 (June 1927), 43-47; and Captain Burdette S. Wright, "Program of the Air Corps Under Administration of Maj. Gen. Mason M. Patrick," Slipstream, VIII:12 (Dec. 1927), 11-13, 16, 18, 22-23.

[8] See Gardner W. Carr, Asst. Ch/Prod, Eng. Div, AS, "Organization and Activities of the Engineering Division of the Army Air Service," U.S. Air Service, VI:6, VII:1 & 2 (Jan, Feb, March 1922), cited after this as Carr articles. The chart illustrating Carr's article presumably reflects the effects on the Engineering Division of General Patrick's reorganization. In any case, the basic structural pattern of that chart continued essentially unchanged until the Materiel Division of the Army Air Corps absorbed the Engineering Division in October 1926. The sources cited below refer to General Patrick's reorganization as resulting in only eight sections -- Planning, Technical, Factory, Flying, Procurement, Supply, Patents, and Military -- a number that does not jibe either with Carr's chart or his description of that period. Perhaps the confusion derives from the use of the word "section" which may have been applied loosely.

"Reorganization of the Office of the C. of A.S.," Aviation, XII:2 (Jan. 9, 1922), 42. Maj. T.H. Bane, Ch, Eng. Div, AS, "A Message from the Commanding Officer," Slipstream, III:21 (Dec. 1921), 2. "Reorganization of the Office of the Chief of Air Service," Air Service News Letter, V:45 (Dec. 27, 1921), 4-5. "Chief of Air Service Receives His Wings," Air Service News Letter, VII:13 (July 10, 1923), 3-4. "General Patrick Reappointed," Aviation, XIX:5 (Aug. 3, 1925), 120. Purtee, 126, 135, 145.

[9] "Two Years' Work at McCook Field," Aviation and Aircraft Journal, X:9 (Feb. 1921), 263.

[10] "Two Years' Work at McCook Field," Aviation and Aircraft Journal, X:9 (Feb. 28, 1921), 263-265. Walker-Wickam, 190.

[11] Frey, 190. Walker-Wickam, 202. Col. Thurman H. Bane, Ch, Eng. Div, AS, "A Message from our Commanding Officer," Sliptream, I:7 (Dec. 15, 1919), 3. Maj. Thurman H. Bane, Ch, Eng. Div, AS, "A Message from the Commanding Officer," Slipstream, III:21 (Christmas 1921), 3. "Two Years' Work at McCook Field," Aviation and Aircraft Journal, X:9 (Feb. 28, 1921), 263-265. Carr article. Rpt, C.F. Taylor, Power Plant Section, Eng. Div, AS, "Aircraft Development since the Armistice - Engines," June 7, 1922. Brochure, Visit of House of Representatives Select Committee of Inquiry into Operations of the U.S. Air Services (Lampert Committee), McCook Field, Dayton, Ohio, Sep. 1924, cited after this as McCook Field brochure. Maj. John F. Curry, Ch, Eng. Div, AS, "McCook Field Review," Aviation, XX:2 (11 Jan. 1926), 47.

By 1920, the Equipment Section designed a portable engine cranker for starting aircraft at McCook Field. The starter was adaptable to all types of engines by using suitable faceplate castings. The cranker was driven by an automobile starting motor using a storage battery, and produces enough torque to spin a cold Liberty 12 engine at 40 rpm. The cranker, mounted on a 1/2 ton truck, was driven into a position in front of the aircraft. Then the cranker was swiveled in its universal bowl so that its shaft paralleled the propeller's axis. The automobile release was then set at the starting position, and the engagement lever was pushed forward until the face plate almost touched the propeller. The necessary adjustments of the elevating and transversing mechanisms were then made and the bowl clamped into position. The engagement lever was then pushed forward so that the face plate engaged the propeller hub nuts, and the starting switch was turned on. As soon as the engine started under its own power, the face plate automatically retracted from the propeller hub nuts, leaving the entire starter clear of the propeller and allowing the starter truck to drive away without interference. "New Portable Cranker a Success," Air Service News Letter, IV:13 (Mar 22, 1920), 9.

[12] Blee, 38. McFarland, 309. Frey, 124, 192. Maurer Maurer, "McCook Field, 1917-1927," The Ohio Historical Quarterly, LXVII (Jan. 1958), 24, 26, cited after this as Maurer article.

Purtee, page 102, and Claussen, page 16, give a different figure for McCook Field personnel at the armistice in November 1918, a total of 2,240: experimental engineering personnel numbered 14 officers and 1,335 civilians; production engineering had 26 officers and 330 civilians; and business and military included 18 officers, 250 civilians, and 267 enlisted men.

[13] The quotation is from Cammen article, 444. Purtee, 119.

[14] "Editorial," Slipstream, III:14 (Sep. 1, 1921), 4.

[15] "A Sincere Start Toward Governmental Economy," Slipstream, III:12 (Aug. 1, 1921), 7, 9, 15. Maj. L.W. McIntosh, Actg. Ch, Eng. Div, AS, "Reductions," Slipstream, III:14 (Sep. 1, 1921), 7.

[16] Maj. A.H. Hobley, Asst. Ch, Eng. Div, AS, "Achievements of the Engineering Division for the Calendar Year 1922," Slipstream, IV:1 (Jan. 1923), 3.

[17] Purtee, 128-129, 135.

Col. Thurman H. Bane, Ch, Eng. Div, AS, "Recent Advances in Aviation," Society of Automotive Engineers Transactions, XV (1920), Part II, 63-86. Maj. T.H. Bane, Ch, Eng. Div, AS, "A Message from the Commanding Officer," Slipstream, III:21 (Dec. 1921), 3. Carr article. Rpt, The Engineering Division: A Consideration of Its Status with Respect to the Air Service and the Aeronautical Industry in the United States and Abroad (Misc. Nr. 238), Eng. Div, AS, Nov. 20, 1924, cited after this as Eng. Div. Rpt. Nr. 238. Gross paper.

Although appropriations for the Army Air Service were very modest, synergistic developments in aviation resulted from cooperation with the Navy, NACA, the Bureau of Standards, the Bureau of Mines, the

Forest Products Laboratory, and manufacturers. To a great extent, the Aeronautical Board of the Army and Navy (known as the Joint Army and Navy Board on Aeronautics from its inception in June 1919 through that December) coordinated their work, preventing duplication and promoting cooperation, including questions relating to development of aircraft. To improve such cooperation, the board was revised in 1924 and again in 1927. The Engineering Division provided surplus aeronautical equipment to the Post Office Department, including night-flying equipment that the division had developed. Similarly, engineering support for the Department of Agriculture included aerial patrol of forests and crop spraying. The division exchanged technical information with England and France, provided technical information to other foreign governments through military attaches, received reports from the League of Nations on foreign developments, and sent representatives to other nations to obtain information.

Eng. Div. Rpt. atch. to ltr, Maj. W.G. Kilner, Exec, AS, to Ch, Eng. Div, AS, n.s. [Report on Research and Development Progress], Feb. 18, 1925, in Purtee, Appendix G. "Two Years' Work at McCook Field," Aviation and Aircraft Journal, X:9 (Feb. 28, 1921), 263. Carr article. Maj. A.H. Hobley, Asst. Ch, Eng. Div, AS, "Achievements of the Engineering Division for the Calendar Year 1922," Slipstream, IV:2 (Feb. 1923), 7. "Aeronautical Board Reorganized," Aviation, XXII:17 (April 25, 1927), 852. Aircraft Year Book 1927, 69-70, 324. Aircraft Year Book 1928, 143-144. Wilson, Autobiography, 42-46.

[18] "Recent Separations From the Service," Slipstream, IV:5 (May 1923), 22.

[19] "The Air Service Appropriations Debate," Aviation, XVIII:3 (Jan. 19, 1925), 74-77.

[20] Purtee, 128. Aircraft Year Book 1926, 300-301; 1930, 568; 1939, 517-519. Walker-Wickam, 196. Charles G. Worman, "McCook Field: A Decade of Progress -- Flight Test Center of the 20s," Aerospace Historian (Spring 1970), 15, cited after this as Worman article.

[21] Eng. Div. Rpt. Nr. 238.

[22] "Annual Report of the Chief of Air Corps," Aviation, XXI:23 (Dec. 6, 1926), 950-951.

[23] "The Lampert Committee Report," Aviation, XIX:26 (Dec. 28, 1925), 906-909.

In the 1920s, periodicals invariably cited large expenditures on aviation by foreign governments to argue that the U.S. was lagging. For example, in 1921 England spent $92.3 million and France $58.8 million on military aeronautics. However, the editor of Aviation challenged the extremely low figures usually cited for U.S. expenditures, arguing that reports were inexact because aviation funding was scattered among several agencies. Controversy centered on what costs were to be counted and on the fact that no consolidated aviation budget existed. Admiral Moffett, for example, argued against charging the upkeep of the Navy's aircraft carriers to the Bureau of Aeronautics, and he opposed inclusion of spending on aviation work by the Bureau of Engineering or the Bureau of Construction and Repairs. Unaccountably, General Patrick, while calling for more funding for military aviation, reported in 1924 that the Air Service had not spent $1,399,001.65 of the 1922-1923 appropriation and $762,321 of the 1923-1924 appropriation -- that is, $2 million that he could have spent on aircraft.

In testimony to the Morrow board, appointed by President Coolidge in September 1925, Grover Loening asserted that from 1919 to 1924 the military services spent $473 million on aviation, but that in those five years, only $3.7 million had been spent on design, development, and testing of new types of aircraft for the Army and Navy -- less than one percent of the money congress appropriated. Loening attributed this situation to "waste on dirigible airships," work being done by government maintenance shops that industry could do, and building of aircraft by McCook Field and the Naval Aircraft Factory.

Asserting that the proposed 1926 budget of $65 million concealed another $25 million in appropriations for pay and other expenditures related to aviation, the periodical Aviation called for a consolidated aviation budget:

> The one important necessity just at this time is a clear detailed and consolidated presentation of all the government expenditures available for aviation so that the public many know in advance of enactment just how it is proposed to spend the funds furnished by the taxpayer.

"The Cost of Our Air Services," Aviation, XVI:17 (April 28, 1924), 445. "$67,241,327.95 Spent by U.S. on Air Services in 1923," Aviation, XVI:17 (April 28, 1924), 446. "Where the Money Goes," Aviation, XVI:19 (May 12, 1924), 509-510. "A $6,000,000 Error in Air Service Costs," Aviation, XVII:5 (Aug. 11, 1924), 861-862. "Analysis of the 1926 Aviation Budget," Aviation, XVII:24 (Dec. 15, 1924), 1390-1391. "Army Air Service Costs Once More Aired," Aviation, XVIII:1 (Jan. 5, 1925), 6-8. "The Air Service Appropriations Debate," Aviation, XVIII:3 (Jan. 19, 1925), 74-77. Loening, Takeoff, 175.

Eventually, after a flurry of inquiries, Aviation came up with an estimate of its own for government spending on aviation: a grand total of $558,634,096.51 for fiscal years 1920 through 1925. The editor considered this figure "the most authoritative now available," but cautioned that its "accuracy is limited" in various ways -- implying that the figure was an underestimate. "Estimated Cost of the Air Service of the United States for Six Years," Aviation, XX:4 (Jan. 25, 1926), 108.

24   Purtee, 152-153. "Hail and Farewell," Slipstream, IV:1 (Jan. 1923), 2. Ltr, Maj. Thurman H. Bane, (USA, Retired), to Maj. L.W. McIntosh, Ch, Eng. Div, AS, n.s. [Retirement Thank You], Dec. 19, 1922, in Slipstream, IV:1 (Jan. 1923), 35. "Major Curry Leaves Dayton," Air Corps News Letter, XI:11 (Aug. 30, 1927), 254.

25   "Annual Report of the Chief of Air Service," U.S. Air Service, VIII:12 (Dec. 1923), 33-34.

26   Purtee, 120. Goldberg, 32, indicates that between 1 July 1920 and 30 June 1921, 330 aircraft crashed, killing 69 men and injuring 27 others in a force of less than 900 pilots and observers.

27   Purtee, 129.

28   "Annual Report of the Chief of Air Service," Aviation, XVII:24 (Dec. 15, 1924), 1392-1394.

General Patrick in 1925, recognizing that money was still insufficient for equipping the tactical units -- let alone providing for war reserve, called for continuing experimentation and research until the best types of aircraft could be developed for building in case of war. Goldberg, 32.

29   Maj. J.F. Curry, Ch, Eng. Div, AS, "What McCook Field Means to Aviation," U.S. Air Services, XI:8 (Aug. 1926), 44. Carr article. Eng. Div. Rpt. atch. to ltr, Maj. W.G. Kilner, Exec, AS, to Ch, Eng. Div, AS, n.s. [Report on Research and Development Progress], Feb. 18, 1925, in Purtee, Appendix G.

30   Eng. Div. Rpt. atch. to ltr, Maj. W.G. Kilner, Exec, AS, to Ch, Eng. Div, AS, n.s. [Report on Research and Development Progress], Feb. 18, 1925, in Purtee, Appendix G.

31   Purtee, 134, 136. The quoted paragraph (in Purtee, 134), used the word "annuals" rather than annals.

Patrick's tenure as chief of the Air Service coincided with a time of record making and racing in national and world aviation. See the articles by Maj. A.H. Hobley, Asst. Ch, Eng. Div, AS, "Achievements of the Engineering Division for the Calendar Year 1922," Slipstream, IV:1, 2, 3 (Jan, Feb, and March 1923); and Capt. Burdette S. Wright, "Progress of the Air Corps Under Administration of Maj. Gen. Mason M.

Patrick," <u>Slipstream</u>, VIII:12 (Dec. 1927), 12, 16, 18, 22, 23, and the accompanying article, "Other Accomplishments of the Air Corps," 23, 26.

32 Maj. Thurman H. Bane, Ch, Eng. Div, AS, "Notes from Speech by Major Bane," <u>Slipstream</u>, IV:1 (Jan. 1923), 42.

Bane made the first flight with de Bothezat's helicopter on December 18, 1922, hovering between two and six feet for a minute and 42 seconds. The helicopter flew again at McCook Field in February 1923, but soon afterward General Patrick canceled the project after more than $200,000 had been spent on it. Power for the helicopter came from a 9-cylinder, 200-horsepower LeRhone rotary motor. The 65-foot long machine was turned over to the Engineering Division's Technical Museum at McCook Field. Dr. George de Bothezat, a Russian emigre and chief of the Special Research Section at McCook Field, received a contract in June 1921.

Maj. T.H. Bane, "A Message from the Commanding Officer," <u>Slipstream</u>, III:21 (Dec. 1921), 3. "The Engineering Division Helicopter," <u>Slipstream</u>, IV:1 (Jan. 1923), 6-7. "The Truth About the McCook Field Helicopter," <u>Slipstream</u>, IV:5 (April 1924), 17-18, 23. Capt. Burdette S. Wright, "Progress of the Air Corps Under Administration of Maj. Gen. Mason M. Patrick," <u>Slipstream</u>, VIII:12 (Dec. 1927), 12, 16, 18, 22, 23. Worman article, 14. Bernard J. Termena, Layne B. Peiffer, and H.P. Carlin, <u>Logistics: An Illustrated History of AFLC and Its Antecedents, 1921-1981</u> (WPAFB, OH: AFLC History Office, c.1982), 36.

33 Heron, <u>Piston Engine</u>, 115. Compare Schlaifer, 168-169.

Heron was born in England on May 18, 1891. His father was a businessman turned actor. Heron suffered from tuberculosis of the hip at age 3, and to the end of his life referred to himself as a "cripple." With his father's help, he was able to get an engineering apprenticeship in shipbuilding. In that position, Heron worked on internal combustion engines, learning about them from the mechanic's position through machinist and foundry worker. At night, he continued his education with engineering courses at Goldsmith College. When his health failed, Heron found employment as a draftsman. Just before WWI, Heron became involved in working on aircraft engines, air-cooled engines, and fuels. By 1915, Heron was chief draftsman.

Heron then moved to the Royal Aircraft Factory, serving as an assistant to Professor A.H. Gibson. Here Heron worked with aluminum air-cooled cylinders, becoming chief of the air-cooled section of the office. After the war, Heron took several other positions before leaving England late in 1920 for Toronto. Coming to McCook Field, he was soon hired by Major George E. A. Hallett and E.T. Jones. Heron worked there from April 1921 through 1925, assisting in the development of large radial engines.

A devoted worker and an able engineer, Heron by 1921 developed successful air-cooled cylinders of nearly 6-inch bore based on his work with Gibson and his own improvements at McCook Field. Against the resistance of their chief engineers, who were committed to water cooling, the Curtiss Aeroplane Company and Wright Aeronautical Corporation accepted contracts from the Army to build prototype radial engines with Heron-designed cylinders. The Lawrance and Heron developments were brought together when Heron in 1926 moved with Jones to Wright Aeronautical where Lawrance was president. Their first product, the Wright J-5, was essentially a Lawrance-type engine with Heron-type cylinders. Lindbergh used this successful engine of the 200-horsepower class for his New York to Paris flight, May 20-21, 1927. Many other pioneering flights and a number of early transport aircraft also used the engine. It won the Collier trophy in 1927. Wright Aeronautical also experimented with air-cooled radial engines with cylinders of larger bore than the J-5, but the first really successful engine of this type was Pratt and Whitney's 425-horsepower Wasp of 1927.

Heron returned to work for the government in late 1928, remaining at Wright Field until early 1934. Between 1927-1928 and between February 1934-1946, Heron worked for Ethyl Corporation where he made significant contributions to aircraft fuels and lubricants. He continued to act as a consultant for Ethyl Corporation, and other companies, until he died in 1963. Charles Fayette Taylor, Aircraft Propulsion: A Review of the Evolution of Aircraft Piston Engines (Washington, DC: Smithsonian Institution Press, 1971), 41, 43, 45. Biography, S.D. Heron, Aero Propul. Lab, AFWAL, n.d. [c. 1983]. S.D. Heron, Autobiography, ed. by Robert V. Kerley (unpublished draft, c.1963), 1, 3, 5, 7-79, 156, 186, 282-287, 300-302.

[34] Heron, Piston Engine, 116. See also, Heron, Autobiography, 133-135.

Heron disdained those who placed total faith in theoretical analyses rather than verifying their conclusions with experimental evidence. See, for example, his discussion of Jones' experiments with a geared Liberty engine, experiments which overturned theoretical "proof" that it could not be done. Indeed, pointing to his work with pistons, Heron claimed with his characteristic self-denigration that "At McCook Field, I indulged in a good deal of senseless experimentation." Heron, Autobiography, 127, 129, 131.

The "patient and intelligent application of the 'cut-and-try' system" was also celebrated by C. Fayette Taylor, "History of the Aeronautical Engine" Aviation, XXI:7 (Aug. 16, 1926), 284-286. In this brief overview of propulsion technology from Leonardo da Vinci (died 1512) through the large air-cooled radials, Taylor makes several other judgments about the progress of engine technology by 1926: "the entire absence of change in the basic features of mechanical arrangement or heat cycle"; the "marvelous advance in materials of construction, especially in the alloys of aluminum and the special alloy steels"; the use of propeller reduction gear; and development of the exhaust-driven turbo-supercharger.

On the eve of the move from McCook Field to Wright Field, one writer suggested -- optimistically -- how far the theoretical aspects of aviation technology had progressed in the decade since WWI. Dr. Michael Watter, "Engineering in Aircraft," Part 1, Aero Digest, X:2 (Feb. 1927), 90, 160, pointing to the knowledge and experience accumulated since the pioneering days of flying, claimed:

> As a result we now possess methods enabling us to cope with the most complicated problems of design and to obtain results as accurate as can be desired for practical purposes. At present we have theoretical methods of calculations and means of scientific testing, and also a library rich with accumulated information of a most varied nature. No matter how original may be an individual design, there are methods of verifying and ascertaining its peculiarities even before the contemplated craft is built.

> Aircraft building is now an established industry, and aircraft design is a science -- not art. It is not necessary to wait until the aircraft is tested to know whether it can fly and how well it flies. Quietly sitting in his office, an engineer is able to predict, with as much certainty as in any other branch of engineering, the performance, capacity and strength of his design. The "cut and try" methods are substituted by careful calculations, and points of importance are verified by tests of details.

Nevertheless, Dr. Watter conceded the practical contributions of the Army, Navy, and NACA to American aircraft engineering. His article, Dr. Michael Watter, "Engineering in Aircraft," Part 2, Aero Digest, X:3 (March 1927), 174, 252-253, which described the typical process of aircraft design, also emphasized the importance of experiential knowledge for aviation technology:

Considering the comparative youth of aircraft engineering one may feel confident by the advance of our knowledge in such a short time. Theory and practice were usually considered antagonistic and it was for aircraft engineering to prove that scrupulous investigation with proper inspection of production and selection of material gives results in perfect agreement with science. Aircraft engineering is more scientific in nature than other branches of engineering and the more scientific it becomes the greater strides will be made in achievement. At present calculations are not considered as a lot of figures but are a reliable source of information, saving cost, eliminating waste and have a real practical value accepted by most "practical men" in the game. Of course to achieve results there must be a harmony between theory and practice and an engineer should deal not only with abstract figures but consider the production difficulties and change in material due to manufacturing processes, etc.

[35] Hallett, who had been commander of the acceptance park in charge of testing at Wilbur Wright Field during WWI, transferred after the armistice to McCook Field. He was chief of the Power Plant Section for four years, leaving in December 1922 to take a position with the General Motors Corporation in Dayton. For Hallett, see "Whooz Who," Slipstream, II:1 (May 15, 1920), 6, 8; and Slipstream, IV:1 (Jan. 1923), 33. See also, "Personals," Air Service Journal, III:20 (Nov. 14, 1918), 13. "Personals," Air Service Journal, IV:2 (Jan. 11, 1919), 28.

Hallett described the Engineering Division's process for developing engines as it had evolved by 1922 -- including contracting procedures, acceptance testing, 50-hour tests, and flight and service testing -- for the 16 types of Air Service aircraft (in six broad categories: pursuit, attack, observation, bombardment, training, and special) in Capt. George E.A. Hallett, Ch, Power Plant Sect, Eng. Div, AS, "A Method of Developing Aircraft Engines," Journal of the Society of Automotive Engineers, X:6 (June 1922), 457-462. Extensive discussion of these 16 types of aircraft and their engines as they had evolved by mid-1923 was provided in "Aircraft Development Since the Armistice," Aviation, XV:1 (July 2, 1923), 6-9.

[36] "Power Plant Section," Slipstream, II:1 (May 15, 1920), 1, 11. The photograph cited did not include the Propeller Branch, of course, which was organizationally part of the Aircraft Section at McCook Field and during the early history of Wright Field.

Iskander Hourwich and W. John Foster published an Air Service Engine Handbook in 1925. Heron, Autobiography, 98, 393.

[37] "Power Plant Section," Slipstream, II:1 (May 15, 1920), 1. See also, Glenn D. Angle, "A New Process of Steel Cylinder Construction," Aviation, XV:27 (Dec. 31, 1923), 794-795; and "A New Engine for Light Planes," Aviation, XVI:6 (Feb. 11, 1924), 146-147.

Glenn D. Angle published Engine Dynamics and Crankshaft Design (Detroit, Michigan: Airplane Engine Encyclopedia Co., 1925.)

[38] "Engineering Branch of Shop Engineering Section," Slipstream, I:3 (15 Sep. 1919), 32. "A Universal Test Engine," Aircraft Journal, IV:n.nr. (27 Dec. 1919), 10.

[39] Glenn D. Angle, "Universal Test Engine," Slipstream, I:7 (15 Dec. 1919), 14-17, 33-34, published also in Aviation, VIII:6 (15 April 1920), 230-234. Slipstream (15 May 1920), 2. Col. Thurman H. Bane, Ch, Eng. Div, AS, "Recent Advances in Aviation," Society of Automotive Engineers Transactions, XV (1920), Part II: 63-86. Maj. T.H. Bane, Ch, Eng. Div, AS, "A Message from the Commanding Officer," Slipstream, II:12 (Christmas 1920), 1. Col. Thurman H. Bane, Ch, Eng. Div, AS, "Colonel Bane on Aircraft Development," Aviation, X:12 (21 March 1921), 360-362. Maj. T.H. Bane, Ch, Eng. Div, "A Message from the Commanding Officer," Slipstream, II:12 (Dec. 1921), 2-3. R.I. Markey, "The Spotlight on McCook Field," The Ohio State Engineer (Nov. 1920), 9-11. "Two Years' Work at McCook Field," Aviation and Aircraft Journal, X:9 (28 Feb. 1921), 263. Rpt, Digest of the Report of Achievements of the Engineering Division, Air Service, for Period 1920-1922, n.d. [c. 1922]. Rpt, C.F. Taylor, Power Plant Sect, Eng. Div, AS, "Aircraft Development Since the Armistice - Engines," 7 June 1922. Maj. A.H. Hobley, Asst. Ch, Eng. Div, AS, "Achievements of the Engineering Division for the Calendar Year 1922," Slipstream, IV:2 (Feb. 1923), 6. Schlaifer, Aircraft Engines, 15-16, 128-131.

40   Maj. A.H. Hobley, Asst. Ch, Eng. Div, AS, "Achievements of the Engineering Division for the Calendar Year 1922," Slipstream, IV:2 (Feb. 1923), 6. Carr article. Rpt, C.F. Taylor, Power Plant Sect, Eng. Div, AS, "Aircraft Development Since the Armistice - Engines," 7 June 1922. Glenn D. Angle, Ch, Eng. Design, Power Plant Sect, Eng. Div, AS, "Progress Toward 1000 Hp. Aircraft Engines," Aviation, XVI:8 (25 Feb. 1924), 198-200. "The Power Plant Section: McCook Field, Ohio," Slipstream (Aug. 1923), 34. Glenn D. Angle, "A New Type of Engine for Large Aircraft," Aviation, XVII:5 (4 Aug. 1924), 832-834. McCook Field brochure. Maj. John F. Curry, Ch, Eng. Div, AS, "McCook Field Accomplishments of Past Year [1924], Aviation Progress (1 Nov. 1925), 4-9. Eng. Div. Rpt. atch. to ltr, Maj. W.G. Kilner, Exec, AS, to Ch, Eng. Div, AS, n.s. [Report on Research and Development Progress], 18 Feb. 1925, in Purtee, Appendix G.

Schlaifer, 15-17, cites the W engine as an object lesson -- a "leading example of direct government development." The engine's welded steel cylinders and W-type arrangement were overcome by the monobloc D-12. A similar fate caused McCook Field to abandon the 24-cylinder, air-cooled X-type engine. Thus, wrote Schlaifer, "After the 1920's the Army made no more attempts to carry out a development completely on its own." He also concluded:

> The fact that even the Army made no attempt after the 1920's actually to carry out its own developments is adequate proof that the services themselves had become convinced that it was virtually impossible to achieve success by this method.

Schlaifer's analysis here, however, ignores several other operative factors affecting McCook Field. The fact is that the Power Plant Section's policy for engine procurement in the early 1920s -- in contrast to the overall policy of the Engineering Division -- did not favor government development and production. In 1923, General Patrick reformed the Air Service's acquisition policy, emphasizing the role of private industry in development and production and shifting the Engineering Division toward evaluation and testing. Schlaifer also does not recognize the severe limits imposed on the Engineering Division's capability in this period by reduced manpower and funding. Manufacturing by the military services, as briefly attempted by the Naval Aircraft Factory, was inevitably a dead-end issue given the attitude in the U.S. toward the military between the world wars, the domination of government policy by private enterprise until the Great Depression, and the positive benefits to be gained from competition.

Moreover, Schlaifer himself, page 41, points out that the government directly financed development of aircraft engines in the U.S. during the years immediately after the armistice of 1918 through development contracts awarded for each project. By 1926, most of the development costs of air-cooled engines were paid for by private firms from their production income; however, the firms still sought development contracts as evidence of government interest in procuring the engine in quantity.

41   "Power Plant Section," Slipstream (15 May 1920), 2.

Schlaifer, page 33, describes NACA's work during the interwar period as largely in the field of airframes rather than engines. NACA's first wind tunnel began operating at Langley Field in 1919.

42   "Liberty Motor," Air Service News Letter, III:n.nr. (22 Aug. 1919), 10-11.

43   "Power Plant Section," Slipstream, II:1 (15 May 1920), 2, 21. Rpt, Digest of the Report of Achievements of the Engineering Division, Air Service, for Period 1920-1922, n.d. [c. 1922]. Rpt, C.F. Taylor, Power Plant Sect, Eng. Div, AS, "Aircraft Development Since the Armistice - Engines," 7 June 1922. Glenn D. Angle, "An Effective Method of Cleaning Engine Oil," Aviation, XVII:24 (15 Dec. 1924), 1398-1399.

An outstanding example of how McCook Field's influence expanded into private industry when its engineers moved involved E.T. Jones, chief of the Power Plant Section. Jones, who had been a professor of mechanical engineering at Cornell University before coming to McCook Field, left the field at the end of 1925, to become chief power plant engineer for the Wright Aeronautical Corporation. On 1 January 1926, Jones replaced Charles Fayette Taylor (head of McCook Field's Power Plant Laboratory from 1919 to 1923) who left Wright Aeronautical in September 1925 for MIT. Two members of Jones's staff at McCook Field went with him, Heron and Harold E. Morehouse. Under Jones and Heron, the Wright company abandoned water-cooled engines and in 1927 they brought the Whirlwind engine to J-5 status. The two men used cylinders based on the type K developed by Heron at McCook Field in 1922, and they improved exhaust-valve cooling using internal cooling by a mixture of sodium and potassium nitrates, also developed by Heron at McCook in 1922. The improvements permitted all the ocean crossings and long-distance flights for which 1927 became famous. The 435-horsepower, 1,652-cubic inch, P-2 Cyclone also was improved and developed into one of the nation's leading air-cooled radials, the R-1750, using a cylinder based on Heron's type M and internally cooled valves. This larger Cyclone tested at 500 horsepower. Jones returned to McCook Field in mid-1927. He died in 1929.

Jones attended public schools in White Plains, New York and received a degree in mechanical engineering from Cornell University in 1911. After that, he was active in aeronautical engine work with Thomas Morse Aircraft Corporation of Ithaca, New York, with the Aviation Section of the Signal Corps just before WWI, as a lieutenant with the Air Service during the war, as a power plant engineer at McCook Field from 1919 to 1922, and as chief of the Power Plant Section from 1923 through 1925. "McCook Men Join Wright," Slipstream, VII:4 (April 1926), 21. Hugo T. Byttebier, The Curtiss D-12 Aero Engine (Washington, DC: Smithsonian Institution Press, 1972), 81. Heron, Autobiography, 101, 117, 139-143, 148-151. Eng. Div. Rpt. Nr. 238. Schlaifer, 191-192. See also, Edward T. Jones, Robert Insley, et al, Aircraft Power Plants (New York: Ronald Press Co., 1926.)

Morehouse, who worked in the Engine Design Branch at McCook Field, designed an 80-cubic inch, two-cylinder, four-cycle, air-cooled engine for use in light aircraft. Dynamometer testing at McCook Field showed a smooth-running engine. Wright Aeronautical Corporation manufactured some of the engines in 1926. Harold E. Morehouse, "The Morehouse Light Plane Engine," and Edmund T. Allen, "The Morehouse Light Plane Engine," Aviation, XVIII:22 (1 June 1925), 602-603, and 609. "Wright-Morehouse Light Engine," Aero Digest, VIII:2 (Feb. 1926), 72-73. Guy Vaughan, "The Wright-Morehouse 25-30 hp. Airplane Engine," Aviation, XXI:8 (23 Aug. 1926), 329-331.

At MIT in 1926, Professor Charles Fayette Taylor foresaw the potential of jet propulsion, judging that water-cooled engines were reaching the limits of their development. Taylor thought that if the gas turbine could be made practical, "it would seem to possess fundamental advantages in respect to mechanical simplicity and the ability to use fuel of low volatility." Taylor in the U.S., Frank Whittle in England, Hans von Ohain in Germany, and Secundo Campini in Italy pushed jet propulsion in the 1930s. Taylor's efforts were not as successful as those in England, Germany, and Italy. Indeed, a Navy report in January 1941 concluded that gas turbine propulsion was not possible for aircraft -- even though in August 1939 the world's first turbojet flight occurred in Germany with the Heinkel He-178. The U.S. flew its first turbojet aircraft, Bell's XP-59 Airacomet, in October 1942 using two engines of British design. Richard P. Hallion, Legacy of Flight: The Guggenheim Contribution to American Aviation (Seattle: University of Washington Press, 1977), 57. Earlier discussion of progress in turbine technology was in "The Internal Combustion Turbine," Aviation, IV:8 (15 May 1918), 529-530, an article taken from Flying, published in London.

Reference to NACA's Technical Report 159 on jet propulsion, and discussion of the theoretical import of gas turbine technology, was also provided by Dr. Michael Watter, "Power and Propulsion at Altitudes," Aero Digest, X:5 (May 1927), 378-379, 483.

44 "Power Plant Section," Slipstream, II:1 (15 May 1920), 21. Lt. Cmdr. E.E. Wilson, USN, "Aircraft Engine Progress," U.S. Air Services, X:2 (Feb. 1925), 15-19.

The period from 1920 to 1925, called a "golden era" of records for the U.S. by Grover Loening, was a period of record-breaking and racing flights. Encouraged by General Billy Mitchell, racing aircraft aimed higher, faster, and farther. Such flights were much more than publicity stunts since they also tested the pilots, their aircraft, and equipment. The data thus garnered advanced technology and safety. Invariably, the Liberty engine played a key role in these flights. After 1926, however, the U.S. turned generally to commercial ventures rather than record flights. Indeed, as Commander Wilson, a Navy enthusiast of air-cooled engines pointed out in his evaluation of record-making flights in 1927:

> To summarize these results, it is apparent at a glance that five World records have returned to the United States in the last few months, that all of these have been obtained with modern air-cooled engines, and that more are in sight. Still more important is the fact that none of the airplanes have been especially designed for the purpose but all have been adapted by making such minor changes as were necessary. When World records are established by service types, it is a good indication that we are entering into a new era in aeronautics.

Loening, Takeoff, 153-154, 191. Cmdr. E.E. Wilson, USN, "The Recent Air Records," Aviation, XXII:22 (30 May 1927), 1129-1130. Maurer article, 31-32. Worman article, 15.

Major Schroeder (affectionately known as "Shorty" because of his 6-feet 4-inch height), chief of the Flight Test Branch at McCook Field, set the world altitude records for one man and for two men. On 6 September 1919, he reached 28,250 feet for a two-man record. On 27 February 1920, Schroeder set the record for one man in a LePere biplane with a Liberty engine and a Moss supercharger, reaching 33,113 feet, a flight that almost ended his life when he fell unconscious for lack of oxygen and spun out of control some 30,000 feet before recovering.

On September 28, 1921, Lieutenant Macready reached 34,509.5 feet over Dayton with a LePere and an improved supercharger; the aircraft also had a specially designed Engineering Division propeller. Macready reached 31,540 feet for a two-man record on May 2, 1924. He reached 37,569 feet himself on January 24, 1925, and 38,704 feet in January 1926.

On December 8, 1921, Lieutenant Wade took a supercharged Martin bomber to 27,120 feet -- almost twice its 14,000-foot regular ceiling -- before a malfunction forced him to descend; he established a record for a two-engine aircraft.

Lieutenants Macready and Oakley G. Kelly, made an endurance flight of 36 hours, 4 minutes, and 30 seconds over Dayton on April 16-17, 1923 in a T-2 monoplane with a Liberty engine and a McCook Field propeller. They then used the same propeller on their transcontinental flight from Mitchel Field, New York, to Rockwell Field, California, on May 2-3, 1923, a distance of 2,520 miles in 26 hours, 50 minutes, and 38 seconds.

"Whooz Who," Slipstream, II:2 (30 June 1920), 3, 5. T.C. McMahon, "High Flying," Slipstream, III:21 (Christmas 1921), 24-25. "Army-Fokker Crosses Continent in 27 Hours," Aviation, XIV:20 (14 May 1923), 524-526. "World's Duration Record Captured," Air Service News Letter, VII:10 (15 May 1923), 1-4. "Accomplishing the Impossible," Air Service News Letter, VII:11 (4 June 1923), 1-5. "Propellers: A Description of the Propeller Branch, McCook Field," Slipstream IV:8 (Aug. 1923), 11. "The Transcontinental Non-Stop Flight," Slipstream, IV:8A (3 Sep. 1923), 12-13. "The Transcontinental Flight: Kelly-Macready and the Famous T-2," Slipstream V:10 (2, 3, 4 Oct. 1924), 118. Termena, Peiffer, and Carlin, 38. Loening, Takeoff, 155.

[45]   "Power Plant Section," Slipstream, II:1 (15 May 1920), 21.

[46]   "Power Plant Section," Slipstream, II:1 (15 May 1920), 21.

47 "The Power Plant Section: McCook Field, Dayton, Ohio," <u>Slipstream</u>, IV:8 (Aug. 1923), 32-33. Rpt, Brig. Gen. William Mitchell, Asst. Ch, AS, subj: Inspection of Air Service Engineering Division, McCook Field, 12 April 1923.

Jones apparently followed Hallett as chief of the Power Plant Section. However, a brief reference to Captain L.L. Snow, who headed the Cooling Systems Branch, suggests that he may have served briefly as chief -- or perhaps as acting chief -- either when Hallett left or during Hallett's tenure. Item on Captain Snow in <u>Slipstream</u>, IV:7 (July 1923), 19.

Similarly, Jones was probably succeeded as chief of the Power Plant Section from January to September 1926 by Lieutenant Carl (perhaps Carlyle) H. Ridenour. One reference even cites Ridenour as chief of the Power Plant Section in September 1923 (perhaps at that time he was only acting chief), when he approved a report prepared by Heron on exhaust valves for aircraft engines. When Ridenour transferred to Hawaii in September 1926, he was succeeded by Captain Theos E. Tillinghast. Tillinghast remained chief of what became the Power Plant Branch until 1930. Heron's report was cited in his <u>Autobiography</u>. "McCook Field, Dayton, Ohio, May 26th," <u>Air Service News Letter</u>, X:9 (8 June 1926), 24.

The difficulty in pinning down whether Captain Snow and Lieutenant Ridenour actually headed the Power Plant Section at McCook Field no doubt results in part from the informal appointment process that prevailed even at Wright Field as late as the 1930s. For example, Daniel Adam Dickey described during an oral history interview how he was replaced as chief of the Propeller Laboratory in these terms:

> <u>Mrs. Walker</u>: *What year was that [when you became chief of the Propeller Laboratory]?*
>
> <u>Mr. Dickey</u>: *I don't know. I can't tell you right off, but I think it was around 1930. I could look it up. We never worried too much about that because, as I say, the military and the civilians worked so close together that a lot of those changes were more or less informal. Lt. Orval Cook [later major general] sat right across from me at the same big desk at Wright Field. He asked me, "Didn't Major Taylor tell you that he has appointed me as Chief?" Because I was going ahead and signing stuff as the Chief. I told him no, he hadn't, so I went to Taylor and, in fact, bawled him out for not telling me first what he was going to do.*

Intvw, Daniel Adam Dickey by Lois E. Walker, Ch, Hist. Ofc, WPAFB, Oh, subj: Oral History, Aug. 1986.

48 "The Power Plant Section: McCook Field, Dayton, Ohio," <u>Slipstream</u>, IV:8 (Aug. 1923), 32. Maj. A.H. Hobley, Asst. Ch, Eng. Div, AS, "Achievements of the Engineering Division for the Calendar Year 1922," <u>Slipstream</u>, IV:2 (Feb. 1923), 7.

Discussion of ignition, including fuels and spark plugs, was provided by Capt. George E.A. Hallett, Ch, Power Plant Sect, Eng. Div, AS, "Ignition from the Engineman's Viewpoint," <u>Aviation</u>, IX:11 (29 Nov. 1920), 354-356.

49 "The Power Plant Section: McCook Field, Dayton, Ohio," <u>Slipstream</u>, IV:8 (Aug. 1923), 32.

50 "The Power Plant Section: McCook Field, Dayton, Ohio," <u>Slipstream</u>, IV:8 (Aug. 1923), 32.

51 "The Power Plant Section: McCook Field, Dayton, Ohio," <u>Slipstream</u>, IV:8 (Aug. 1923), 32-33.

52 "The Power Plant Section: McCook Field, Dayton, Ohio," <u>Slipstream</u>, IV:8 (Aug. 1923), 33.

53   "The Power Plant Section: McCook Field, Dayton, Ohio," <u>Slipstream</u>, IV:8 (Aug. 1923), 33.

54   "The Power Plant Section: McCook Field, Dayton, Ohio," <u>Slipstream</u>, IV:8 (Aug. 1923), 33.

An aircraft without an engine was flown at McCook Field in 1923 by W.F. (Fred) Gerhardt. The seven-wing Gerhardt Cycleplane left the ground on several occasions and remained airborne for brief periods solely by the pumping power of the pilot. The project, which dated back to 1920, originated in an argument among engineer friends at McCook Field. After submitting their arguments to mathematical analysis, a theoretical solution seemed possible. Thus, in February 1923, Gerhardt and Pratt made a layout. The two men got the assistance of Bob Anderson who made the tractor propeller. The first design included a prone pilot, but pumping in that position was too difficult. Therefore, they changed the pilot's position to a bicycle stance. They built a model by July 1923, "through the kind co-operation of the Engineering Division officials, who were curiously, if not seriously concerned with the strange craft." The men used the helicopter hangar to assemble and test the cycleplane.

The more of it that was assembled the more discouraged we became of ever making such a flimsy looking structure hold together long enough to be wheeled outside the hangar, let alone take the air with a load of 250 pounds.

But they took the craft out of the hangar at night on Friday July 13, 1923. Despite their efforts to push the cycleplane into motion, the craft did not move. Four days later, the men tried again, this time using an automobile to tow the craft. It managed to leave the ground without a pilot. Next, Pratt

was able with considerable exertion to keep the machine aloft for a considerable period of time without further towing power from the automobile. This in itself, as we later realized, was actual flight by man power.

After several more days, they resumed testing, both towing the vehicle and with the operator, achieving a total horsepower drag of 0.8 (which was the main data that they were seeking.) Two days later, they towed the craft again and after takeoff, "wound up the propeller." After experimenting with three pitch settings, the spring balance read zero while the craft was off the ground for several inches over a distance of 20 feet -- "proof that steady level flight was accomplished." In October 1923, the craft once again was towed and the operator began pumping the pedals,

but no sooner had the craft cleared the ground than something essential to the rigidity of the structure gave way and the fuselage gently folded in two and the wings collapsed backward. It must be admitted that this method of disassembling was not premeditated, and is not to be recommended except for its dispatch.

Fred W. Gerhardt, "Flight by Man Power," <u>Slipstream</u>, V:1 & 2 (Feb. 1924), 9-11, 36.

55   Carr article. Lester D. Seymour, "The Army Air Service and the Liberty Engine," <u>Air Service News Letter</u>, VI:29 (19 Oct. 1922), 1-3. Ltr, E.T. Jones, Ch, Power Plant Sect, Eng. Div, AS, to Ch, Eng. Div, AS, subj: Data on Power Plant Development, 13 Jan. 1925, <u>in</u> Purtee, Appendix F. Eng. Div. Rpt. atch. to ltr, Maj. W.G. Kilner, Exec, AS, to Ch, Eng. Div, AS, n.s. [Report on Research and Development Progress], 18 Feb. 1925, <u>in</u> Purtee, Appendix G. <u>Slipstream</u> (Aug. 1923), 33-34. Capt. Burdette S. Wright, "Progress of the Air Corps Under Administration of Maj. Gen. Mason M. Patrick," <u>Slipstream</u>, VIII:12 (Dec. 1927), 16.

56   Maj. T.H. Bane, Ch, Eng. Div, AS, "A Message from the Commanding Officer," <u>Slipstream</u>, III:21 (Dec. 1921), 2. Maj. A.H. Hobley, Asst. Ch, Eng. Div, AS, "Achievements of the Engineering Division for the Calendar Year 1922," <u>Slipstream</u>, IV:2 (Feb. 1923), 6. Rpt, Digest of the Report of Achievements of the Engineering Division, Air Service, for Period 1920-1922, n.d. [c. 1922]. "The Power Plant Section:

McCook Field, Dayton, Ohio," Slipstream, IV:8 (Aug. 1923), 34. McCook Field brochure. "Almen 'Barrel' Engine Slow Process," Slipstream, VII:1 (Jan. 1926), 29. Ltr, E.T. Jones, Ch, Power Plant Sect, Eng. Div, AS, to Ch, Eng. Div, AS, subj: Data on Power Plant Development, 13 Jan. 1925, in Purtee, Appendix F. Eng. Div. Rpt. atch. to ltr, Maj. W.G. Kilner, Exec, AS, to Ch, Eng. Div, AS, n.s. [Report on Research and Development Progress], 18 Feb. 1925, in Purtee, Appendix G. Maj. John F. Curry, Ch, Eng. Div, AS, "McCook Field Accomplishments of Past Year [1924], Aviation Progress (1 Nov. 1925), 4-9. "Great Improvements Shown in Engines," U.S. Air Services, XII:1 (Jan. 1927), 44. Heron, Piston Engine, 100-101. Taylor, 57.

57   Heron, Piston Engine, 17-18, 41. Byttebier, 43-44, 51-54, 64-67, 71-73. Lt. Cmdr. E.E. Wilson, USN, "Aircraft Engine Progress," U.S. Air Services, X:2 (Feb. 1925), 15-19.

The Army received two of the first D-12 engines produced and put them to test as high speed engines. The Army ordered two high-speed aircraft from Curtiss to compete as R-6s in the 1922 Pulitzer races held in Detroit in October. Curtiss engineers designed the aircraft to reduce drag, incorporating the radiators as part of the wing surfaces. The R-6 racers were fully braced biplanes with fixed undercarriages. The race was a sweeping victory for the Army and the D-12 engine which took the first four places. The two R-6 racers came in first and second followed by two Curtiss Navy racers, surpassing other aircraft with more powerful Wright T-2 and Packard 1A-2025 engines. The winner, Lieutenant Russell Maughan, averaged 206 miles per hour, 30 mph more than the 1921 winner. A few days after the race, General Billy Mitchell used the Pulitzer winner to bring the world's speed record to the U.S. for the first time with a mean speed over four runs of 223 miles per hour.

Early in 1923, the French regained the speed record at 233 mph using a Wright H-3 engine, but they kept it only a few weeks. On 29 March 1923, Lieutenant Maughan established a new world's record of 244.9 mph with a biplane R-6 racer powered by the D-12 with new Stromberg NA-Y5 carburetors and a new all-metal Reed propeller. On 23 June 1924, Lieutenant Maughan flew a PW-8 aircraft with a Reed propeller and a D-12 engine, improved based on results of testing at McCook Field in 1923 and 1924, from Mitchel Field, Long Island, to Crissy Field, San Francisco, in 21 hours and 48 minutes. Actual flying time for the 2,645 miles was 18 hours and 20 minutes (compared to the 26 hours, nonstop, of the Liberty-powered T-2 in May 1923.) Maughn's average flying speed was over 156 mph.

In 1924, with the establishment of the D-12 as a military engine, interest in racing (regarded as the peacetime substitute for war) began to wane. The D-12 engine was dominant in the 1924 Pulitzer race in Dayton. The two R-6 aircraft and the winning R-3 monoplane had 500-horsepower D-12A engines. But the rules of the race required the R-6 racers to use wooden propellers, a tragic mistake. The propeller on one of the Curtiss racers burst during a dive to gain speed. After that accident, wooden propellers gave way to the new metal propellers.

A total of 1,192 D-12 engines was built during the period from 1922-1932. The price of a standard D-12 during the 1920s was about $9,100, and the time between overhauls was about 200 hours. After 1930, the D-12s were rapidly displaced by engines of newer and different designs. Byttebier, 54-60, 65-66, 69, 74. See also, Lt. S.H. Wooster, USN, "The Case for the Metal Propeller," U.S. Air Services, X:5 (May 1925), 24-27.

58   Joseph Lawrence Nayler, Aviation: Its Technical Development (Philadelphia: Dufour Editions, 1965), 147-148. Byttebier, 67-71. Schlaifer, 172-173. "The Curtiss D-12 Engine in Gt. Britain," Aviation, XX:25 (21 June 1926), 945.

59   Byttebier, 68-69, 82, 100, 103-104. Schlaifer, 28. Archibald Black, "Developing a Thorobred," Aero Digest, VII:4 (Oct. 1925), 524-527, 570. Rpt, Brig. Gen. William E. Gillmore, Ch, Matl. Div, AAC, subj: Second Annual Report of the Chief, Materiel Division, Air Corps, Fiscal Year 1928 [1 July 1927-30 June 1928].

60   "The Power Plant Section: McCook Field, Dayton, Ohio," Slipstream, IV:8 (Aug. 1923), 34.

61   "The Power Plant Section: McCook Field, Dayton, Ohio," Slipstream, IV:8 (Aug. 1923), 35. Rpt, C.F. Taylor, Power Plant Sect, Eng. Div, AS, "Aircraft Development Since the Armistice - Engines," 7 June 1922. Maj. A.H. Hobley, Asst. Ch, Eng. Div, AS, "Achievements of the Engineering Division for the Calendar Year 1922," Slipstream, IV:2 (Feb. 1923), 6-7. Eng. Div. Rpt. atch. to ltr, Maj. W.G. Kilner, Exec, AS, to Ch, Eng. Div, AS, n.s. [Report on Research and Development Progress], 18 Feb. 1925, in Purtee, Appendix G. Heron, Piston Engine, 50-52, 56. Taylor, 78-79. Heron, Autobiography, 127, 129.

Comparative results of testing geared and direct-drive propellers by the Navy and McCook Field were described by Richard M. Mock, "Geared Down Propellers and the Efficiency of Commercial Airplanes," Aviation, XXII:22 (30 May 1927), 1137-1140. Geared propellers improved performance (takeoff, climb, and speed) but increased weight, complicated design, and added cost.

62   "The Power Plant Section: McCook Field, Dayton, Ohio," Slipstream, IV:8 (Aug. 1923), 35. "The Allison Four-Engine Transmission," Aviation, XIX:18 (2 Nov. 1925), 640.

At the direction of General Billy Mitchell, the Engineering Division held a competition for a large bomber. A design submitted by Walter H. Barling won, and Witteman Aircraft Corporation built the bomber for the Air Service. Barling was a passenger during its first flight on 23 August 1923 at Wilbur Wright Field. The 40,000-pound bomber stayed in the air for 28 minutes reaching 2,500 feet and 93 miles per hour. Eventually, a special hangar was built at the field to house the aircraft which was 28 feet high, 65 feet long, and had a wing span of 120 feet. When loaded, the bomber, even with six 12-cylinder, 400-horsepower Liberty engines, was underpowered and could not cross the Appalachian mountains safely. Top speed with a full load of bombs was 95 miles per hour. By 1925, the Engineering Division classified this heavy bomber as "obsolescent." Interestingly, Barling regarded his aircraft as "small," predicting that in the future "large" aircraft would be built.

Walter Barling, "The Barling Bomber," Slipstream, IV:2 (Feb. 1923), 8, 18. "The Barling Bomber: Largest Airplane in the World," Slipstream, V:10 (2, 3, 4 Oct. 1924), 55. McCook Field brochure. Eng. Div. Rpt. atch. to ltr, Maj. W.G. Kilner, Exec, AS, to Ch, Eng. Div, AS, n.s. [Report on Research and Development Progress], 18 Feb. 1925, in Purtee, Appendix G. Maurer article, 30-31. Boyne, Wings, 12.

63   "The Power Plant Section: McCook Field, Dayton, Ohio," Slipstream, IV:8 (Aug. 1923), 35. Maj. A.H. Hobley, Asst. Ch, Eng. Div, AS, "Achievements of the Engineering Division for the Calendar Year 1922," Slipstream, IV:2 (Feb. 1923), 7. Carr article.

64   "The Power Plant Section: McCook Field, Dayton, Ohio," Slipstream, IV:8 (Aug. 1923), 35.

65   "The Power Plant Section: McCook Field, Dayton, Ohio," Slipstream, IV:8 (Aug. 1923), 35. Heron, Piston Engine, 89. Worman article, 15. Nayler, 151. Taylor, 55-56. Rpt, Brig. Gen. William E. Gillmore, Ch, Matl. Div, AAC, subj: Second Annual Report of the Chief, Materiel Division, Air Corps, Fiscal Year 1928 [1 July 1927-30 June 1928].

Heron believed that after his initial testing of ethylene glycol, the Power Plant Section did not work on its application again until Captain Tillinghast, then chief of the Power Plant Branch, resumed work with it late in 1928 or early 1929 as a high-temperature coolant. In fact, however, the Materiel Division was already recommending its procurement by the end of fiscal year 1928 (June 1928.) Heron, Autobiography, 98.

When Prestone was introduced as an automobile antifreeze, a rust inhibitor and a leak sealer were added to its ethylene-glycol base. Prestone then became unsuitable for aircraft use because the added chemicals turned to jelly when the solution exceeded 250 degrees Fahrenheit. Byttebier, 78-79.

66 Eng. Div. Rpt. Nr. 238.

67 A.M. Jacobs, "General Gillmore Comes to Dayton," <u>Air Corps News Letter</u>, X:16 (13 Dec. 1926), 14-15. Brig. Gen. William E. Gillmore, Ch, Matl. Div, AAC, "Review of Year's Developments in the Army Air Corps," <u>Slipstream</u>, VIII:3 (March 1927), 7. <u>Aircraft Year Book 1927</u>, 306-307. <u>Who's Who in American Aeronautics</u> (New York City: The Gardner Publishing Co., Inc., 1925.)

Schlaifer, pages 38-39, cites MacDill as an example of how the U.S. failed to use technical talent to its fullest advantage. MacDill, chief engineer in 1923, went to Command and General Staff School in 1924-1925, and then returned as chief engineer in 1925. A trained engineer with a doctorate from MIT, MacDill filled the position of chief engineer of Wright Field from 1928-1934, then was "advanced" by appointment as executive officer to the chief of the Air Corps.

68 Brig. Gen. William E. Gillmore, Ch, Matl. Div, AAC, "Review of Year's Developments in the Army Air Corps," <u>Slipstream</u>, VIII:3 (March 1927), 7. Capt. Burdette S. Wright, "Progress of the Air Corps Under Administration of Maj. Gen. Mason M. Patrick," <u>Slipstream</u>, VIII:12 (Dec. 1927), 12, 16, 18, 22, 23. <u>Aircraft Year Book 1927</u>, 205-209. <u>Aircraft Year Book 1928</u>, 255, 257, 259-260, 262.

69 Rpt, Maj. Leslie MacDill, Ch, Exp. Eng. Sect, Matl. Div, AAC, subj: Status of Experimental and Service Test Projects, Engineering Program - Fiscal Year 1927, As of 1 January 1927, 10 Jan. 1927. Brig. Gen. William E. Gillmore, Ch, Matl. Div, AAC, "Review of Year's Developments in the Army Air Corps," <u>Slipstream</u>, VIII:3 (March 1927), 12. Rpts, Brig. Gen. William E. Gillmore, Ch, Matl. Div, AAC, subj: First and Second Annual Reports of the Chief, Materiel Division, Air Corps, Fiscal Year 1927 [1 July 1926-30 June 1927] & Fiscal Year 1928 [1 July 1927-30 June 1928]. McCook Field brochure. Eng. Div. Rpt. atch. to ltr, Maj. W.G. Kilner, Exec, AS, to Ch, Eng. Div, AS, n.s. [Report on Research and Development Progress], 18 Feb. 1925, <u>in</u> Purtee, Appendix G. Lt. Cmdr. E.E. Wilson, USN, "Aircraft Engine Progress," <u>U.S. Air Services</u>, X:2 (Feb. 1925), 15-19. "Review of Curtiss Aeroplane & Motor Co., Inc.," <u>Slipstream</u>, VIII:4 (April 1927), 20, 22, 24, 27. <u>Aircraft Year Book 1927</u>, 211, 229. "Recent Developments in Aero Engines," <u>Slipstream</u>, VI:5 (May 1925), 19. Col. J.G. Vincent, "Designing Engines Into Airplanes," <u>Aviation</u>, XIX:22 (30 Nov. 1925), 776-778. Maj. John F. Curry, Ch, Eng. Div, AS, "McCook Field Review," <u>Aviation</u>, XX:2 (11 Jan. 1926), 48. "Great Improvements Shown in Engines," <u>U.S. Air Services</u>, XII:1 (Jan. 1927), 44. Arthur Nutt, "The Curtiss V-1550 and GV-1550 Engines," <u>Aviation</u>, XXII:10 (7 March 1927), 465-469. "Curtiss Water-Cooled Engines Shine at Spokane," <u>U.S. Air Services</u>, XII:11 (Nov. 1927), 38. Heron, <u>Piston Engine</u>, 41.

70 Brig. Gen. William E. Gillmore, Ch, Matl. Div, AAC, "Review of Year's Developments in the Army Air Corps," <u>Slipstream</u>, VIII:3 (March 1927), 12. Rpt, Brig. Gen. William E. Gillmore, Ch, Matl. Div, AAC, subj: First Annual Report of the Chief, Materiel Division, Air Corps, Fiscal Year 1927 [1 July 1926-30 June 1927]. McCook Field brochure. Eng. Div. Rpt. atch. to ltr, Maj. W.G. Kilner, Exec, AS, to Ch, Eng. Div, AS, n.s. [Report on Research and Development Progress], 18 Feb. 1925, <u>in</u> Purtee, Appendix G. <u>Aircraft Year Book 1927</u>, 229-230.

Packard also developed a diesel engine, designed by L.M. Woolson, which was certified by the Civil Aeronautics Administration in 1930. Packard's 9-cylinder radial air-cooled diesel, the first diesel engine to power an airplane, set the world's nonrefueling duration record for heavier-than-air craft between 25-May 28, 1931. The engine, its designer, and manufacturer received the Collier trophy for 1931, but the engine failed to become an important aircraft power plant. By the beginning of WWII, with the general use of high-octane fuels, diesel engines could not compete with the conventional spark-ignition type, and its development ended. NACA's research work on diesels for aircraft during the late 1920s and early 1930s, though extensive, also found no practical application. Taylor, 59-60.

71 As quoted in Eng. Div. Report atch. to ltr, Maj. W.G. Kilner, Exec, AS, to Ch, Eng. Div, AS, n.s. [Report on Research and Development Progress], 18 Feb. 1925, <u>in</u> Purtee, Appendix G. Descriptions of the

Packard models 1500 and 2500 engines were in "The New Packard Aircraft Engines," Aviation, XVIII:19 (11 May 1925), 517-520.

72 McCook Field brochure. "Inverting the Liberty Engine," Air Service News Letter, VIII:10 (31 May 1924), 10-11. "Great Improvements Shown in Engines," U.S. Air Services, XII:1 (Jan. 1927), 46. Byttebier, 72. Schlaifer, 160.

Heckert, who came to McCook Field in 1917 as a power plant engineer, left Wright Field in the 1930s. He was a mechanical engineering graduate of Carnegie Institute of Technology. He died in 1976 at age 86. "Aircraft Pioneer Dies; Developed Engines," Journal Herald, (1 April 1976), 34.

After WWI, the Navy used the Liberty engine on its NC aircraft, built by Curtiss, to make the first flight across the Atlantic Ocean. Three aircraft, the NC-1, NC-3, and NC-4, flew from Long Island on 8 May 1919 to Newfoundland (the NC-2 was dismantled to provide spare parts.) Each NC aircraft was equipped with four Liberty engines, and their engines performed well throughout the trip. On 16 May, they set out for the Azores, but two of the aircraft broke down along the way. The crew of NC-1 taxied for five hours before being rescued by a destroyer; however, the aircraft sank from damage caused by the ship while trying to salvage it. Meanwhile, the NC-3 taxied to the Azores, a 52-hour, 205-mile venture! The NC-4 aircraft, after refueling at sea, continued to the Azores. This aircraft reached Lisbon on 27 May 1919 and Plymouth on 31 May. Cmdr. H.C. Richardson, Ch. Engr, Naval Aircraft Factory, "Some Lessons of the Transatlantic Flight," Aviation, VIII:11 (1 July 1920), 445-446. "Previous Great Trans-Atlantic Flights," Slipstream, VIII:7 (July 1927), 35. Flight (New York: Simon & Schuster, Inc., 1953), 94.

Liberty engines in DH-4s and Hispano-Suiza engines in JN-4Hs performed well in the transcontinental air race set up through the influence of General Billy Mitchell in 1919. The race, a round trip from New York to the Pacific coast and return, covered 5,400 miles. Forty-nine aircraft set out from Mineola, New York, and 14 from San Francisco. The winner took nine days for the round trip, flying some 50 hours, and making stops every 100 to 200 miles. The contest, a grueling test for the Air Service, produced much technical information. Grover Loening, Takeoff Into Greatness (New York: G.P. Putnam's Sons, 1968), 147.

One of the most celebrated uses of the standard Liberty engine was the Douglas Around-the-World flight of 1924, an aviation first which received the Collier trophy and was cited by General Patrick as proof that no country was "immune from attack by aircraft." In January 1923, General Patrick directed a study of the possibilities of an aircraft flight around the world. After the study, Douglas Airplane Company was given a contract to build four single-engine, two-place biplanes modeled after the DT-2 torpedo aircraft manufactured for the Navy. These World Cruisers were constructed with Liberty engines under the direction of the Engineering Division. To power the aircraft, 35 Liberty engines were overhauled, tested, and shipped to stations set up around the world. The engines served as replacements in lieu of overhauling engines en route. Only 20 engines were actually used. In addition, 31 propellers (13 oak for use with pontoons and 18 walnut for use with landing gear) built by the Engineering Division with special brass tips were shipped, mostly to foreign countries along the route. An electric starter and an advanced ignition system, both developed at McCook Field, were also used on the flight. The 175-day, 27,553-mile journey began officially at Seattle, Washington, on 6 April 1924 and ended there on 28 September after a flying time of 363 hours. Two aircraft were lost, but the other two completed the entire journey. The different weather conditions encountered enabled the Air Service to learn much about the 12-volt battery system, the generators, starters, distributors, spark plugs, cooling system, fuel system, and oil system.

"Around the World by Airplane," Slipstream, IV:12 (Dec. 1923), 13-15. "Blazing the Trail for Around-the-World Flyers," Slipstream, V:3 (March 1924), 29-30, 33-34. "The Cruise of the Age," Slipstream, V:4 (April 1924), 15-16, 24-25. "Power for the Around-the-World Flight," Slipstream, V:5 (May 1924), 33-34, 37. "Brief History of the Round the World Flight," Aviation, XVII:12 (22 Sep. 1924), 1015-1018. "Factors of Success in the World Flight," Aviation, XVII:12 (22 Sep. 1924), 1018-1021. "Power for the

Around the World Flight," Aero Digest, V:3 (Sep. 1924), 170-172. "Salient Notes Concerning the Round-the-World Flight," Slipstream, V:9 (Sep. 1924), 85, 87, 89. An issue of Slipstream, V:11 (Nov. 1924), was devoted to the world flight. Maj. John F. Curry, Ch, Eng. Div, AS, "McCook Field Accomplishments of Past Year [1924], Aviation Progress (1 Nov. 1925), 4-9. Maj. Gen. Mason M. Patrick, Ch, AS, "Engineering the Round-the-World Flight," U.S. Air Services, X:6 (June 1925), 21-26. Aircraft Year Book 1925, 66-95. McCook Field brochure. Termena, Peiffer, and Carlin, 6, 38. Loening, Takeoff 157-158. Hallion, Guggenheim Contribution, 15.

One awkward aspect of the world flight was the need to change landing gear for pontoons at different stations. The Engineering Division contracted with Loening in December 1923 for an amphibian aircraft based on his claims that he could build one that could fly as fast as a land plane. In June 1924, the first aircraft was ready. Testing of the production aircraft at McCook Field, which began in February 1925, showed a top speed of 122 miles per hour. Ten more aircraft were ordered. On 21 December 1926, five Loening COA-1 amphibians with the inverted Liberty engine and Standard Steel propellers departed on a good-will tour around South America. The group, led by Major Herbert A. Dargue, started from Kelly Field in San Antonio. The 10 officers did their own maintenance. After departing from San Antonio, Texas, they visited 25 nations and colonies in North, Central, and South America, landing in 100 different cities and towns. In addition to crossing the Andes, they flew long stretches over open seas and tropical jungles. One stretch covered 1,200 miles in a day. They made many landings and takeoffs on both land and water. Near Buenos Aires, two of the aircraft were wrecked in a mid-air collision, and one of the crews was lost. Three of the five aircraft and a replacement plane completed the entire 20,470-mile journey. On 2 May 1927, the four aircraft landed at Bolling Field in Washington, D.C. in time for the opening of the Pan-American Air Commission Conferences and the All America Aircraft Display of the Aeronautical Chamber of Commerce there. President and Mrs. Coolidge, the cabinet, General Patrick, and other dignitaries met the fliers. Loening, Takeoff, 138, 192-194. Loening, Our Wings, 149-153. Lt. W.H. Brookley, AS, "The New Loening Amphibian," Air Service News Letter, IX:7 (20 April 1925), 1-3. "The Loening Amphibian," Slipstream, VIII:1 (Jan. 1927), 14-15, 18-19. "Pan-American Flyers End Their Long Journey," Air Corps News Letter, XI:6 (14 May 1927), 145-146. "Home Coming of Pan-American Fliers," Aviation, XXII:20 (16 May 1927), 1039-1040. "All America Aircraft Display Huge Success," Slipstream, VIII:6 (June 1927), 14-15. "Unlimited Benefits Will Follow Pan-American Flight," U.S. Air Services, XII:6 (June 1927), 25-28. "The Pan-American "Good Will" Flight," Slipstream, VIII:7 (July 1927), 27, 35. Aeronautical Year Book 1928, 36-37.

For the Navy, Loening replaced the Liberty on his amphibian with P&W's Wasp engine in 1927. The substitution worked well. Detailed description of Loening's aircraft was provided in "The Loening OL-8 Amphibian," Aviation, XXIII:17 (24 Oct. 1927), 1002-1008.

The reliability of the regular Liberty engines was excellent. They were used for many record flights as well as in air mail service, on the airways, and for cross-country work. But in 1925, one national aviation periodical labeled the Army's upgrading of Liberty engines a "shortsighted economy" because they were "obsolescent" and "expensive" compared to "development of another modern airplane engine." The Army's alternative, however, was not much better: the Air Service offered Liberty 12 engines for sale, asking $2,000 each plus a handling charge of $10.00 per engine. During fiscal year 1926, 867 new and 926 used Liberty engines were declared surplus; 200 of these were transferred to the Coast Guard for use as marine engines to enforce prohibition; hundreds of others were sold, but 511 new and 500 used Liberty engines were still available at the end of that business year (30 June 1926.) Late in 1926, Secretary of Commerce Herbert C. Hoover publicly expressed his doubts about the advisability of selling surplus Liberty engines for less than the cost of manufacture since the sale was a blow against the nation's aviation interests, undermining development and sale of more modern power plants. "Misplaced Economy," Aviation, XVIII:22 (1 June 1925), 597. "Sale of New Liberty Engines," Aviation, XIX:1 (6 July 1925), 8. "Annual Report of the Chief of Air Corps," Aviation, XXI:23 (6 Dec. 1926), 950-951. "Junk Liberty Engines," Aero Digest, IX:5 (Nov. 1926.)

The inertia caused by World War I equipment, especially engines, was not entirely overcome by the Air Corps until the 1930s. Continued use of engines from WWI hurt the U.S. aircraft engine industry and affected the development and procurement of aircraft. The Lampert committee in December 1925 reported that the Air Service still had about 10,000 surplus Liberty engines. As late as 1930, the Air Corps still had more than $40 million worth of Liberty engines in stock which it intended to use as spares. During the 1920s, the Liberty was required to undergo a complete overhaul after 50 hours of service. In 1930, the Air Corps increased the time between overhauls for the Liberty to 125 hours but limited the number of major overhauls during its service life to four. In 1931, major overhauls of the Liberty engine were forbidden. Congress finally prohibited the production of new aircraft using Liberty engines. Goldberg, 37. Purtee, 120. "The Lampert Committee Report," Aviation, XIX:26 (28 Dec. 1925), 906-909. Termena, Peiffer, and Carlin, 31-32.

Other engines showed remarkable staying power long after WWI, discouraging technological progress. These engines included the 90-horsepower, 8-cylinder, water-cooled Curtiss OX-5 and the Hispano-Suiza. The Hispano-Suiza was still in service in 1927 eight years after production was discontinued by the Wright Martin Aircraft Corporation, predecessor of the Wright Aeronautical Corporation. After the armistice, the U.S. military services had little use for the thousands of engines already produced and commercial aviation was still only just beginning. Surplus engines were sold at low prices to anyone interested. "Ye Good Old Hisso: Famous War-Time Motor Still Giving Service," Slipstream, VIII:6 (June 1927), 10, 34. Loening, Takeoff, 134. Wilson, Autobiography, 16-17. Heron, Piston Engine, 15-16.

[73] Brig. Gen. William E. Gillmore, Ch, Matl. Div, AAC, "Review of Year's Developments in the Army Air Corps," Slipstream, VIII:3 (March 1927), 12-13. Brig. Gen. William E. Gillmore, Ch, Matl. Div., AAC, "The Development of Air-Cooled Engines," Airway Age, IX:7 (July 1928), 13-14. McCook Field brochure. Eng. Div. Rpt. atch. to ltr, Maj. W.G. Kilner, Exec, AS, to Ch, Eng. Div, AS, n.s. [Report on Research and Development Progress], 18 Feb. 1925, in Purtee, Appendix G. "Review of Curtiss Aeroplane & Motor Co., Inc.," Slipstream, VIII:4 (April 1927), 20, 22, 24, 27. Aircraft Year Book 1927, 211, 229. Hobert L. Wilson, "Army Engineering Developments of Interest to Commercial Aviation," Slipstream, VI:8 (Aug. 1925), 9. Maj. John F. Curry, Ch, Eng. Div, AS, "McCook Field Review," Aviation, XX:2 (11 Jan. 1926), 47. Heron, Piston Engine, 27-28, 41. Heron, Autobiography, 103.

[74] Carr article. McCook Field brochure. C. Fayette Taylor, "Carburetors for Aircraft Engines," Aviation, XX:10 (8 March 1926), 326-327. Rpt, Maj. Leslie MacDill, Ch, Exp. Eng. Sect, Matl. Div, AAC, subj: Status of Experimental and Service Test Projects, Engineering Program - Fiscal Year 1927, As of 1 January 1927, 10 Jan. 1927. Brig. Gen. William E. Gillmore, Ch, Matl. Div, AAC, "Review of Year's Developments in the Army Air Corps," Slipstream, VIII:3 (March 1927), 12-13. Rpts, Brig. Gen. William E. Gillmore, Ch, Matl. Div, AAC, subj: First and Second Annual Reports of the Chief, Materiel Division, Air Corps, Fiscal Year 1927 [1 July 1926-30 June 1927] & Fiscal Year 1928 [1 July 1927-30 June 1928]. Ltr, E.T. Jones, Ch, Power Plant Sect, Eng. Div, to Ch, Eng. Div, subj: Data on Power Plant Development, 13 Jan. 1925, in Purtee, Appendix F. Eng. Div. Rpt. atch. to ltr, Maj. W.G. Kilner, Exec, AS, to Ch, Eng. Div, AS, n.s. [Report on Research and Development Progress], 18 Feb. 1925, in Purtee, Appendix G. Aircraft Year Book 1927, 215-216. "The Scintilla Magneto for Aircraft Engines," Aviation, XVIII:11 (16 March 1925), 294-295. Hobert L. Wilson, "Army Engineering Developments of Interest to Commercial Aviation," Slipstream, VI:8 (Aug. 1925), 6, 8-9. "Scintilla Aircraft Magnetos Gain Wide Favor," Aviation, XXII:13 (28 March 1927), 615-616. Wilson, Autobiography, 39-42.

Technical discussion of propulsion equipment developments -- like bearings, lubricating oils, ignitions, spark plugs, and ethylene glycol -- can be found in Heron, Piston Engines.

According to Schlaifer, pages 100-102, improvements in aircraft engine carburetors foundered on the monopoly held by Stromberg Motor Devices Company. The company's float carburetor hindered aircraft maneuvers and was subject to throttle icing. Technical progress was limited even though a floatless automobile carburetor had been developed by the end of the 1920s and the basic patent on the control used in the floatless, nonicing carburetor produced by Stromberg in 1938 existed in the 1920s. Stromberg made no effort to develop a suitable aircraft carburetor until after 1935 when competition by Chandler-

Groves' floatless carburetor began. The competition caused Stromberg to invest capital in development and stimulated engineers to rethink their views. The result was that floatless, nonicing Stromberg pressure carburetors dominated U.S. high-power aircraft engines after 1945.

75 Eng. Div. Rpt. atch. to ltr, Maj. W.G. Kilner, Exec, AS, to Ch, Eng. Div, AS, n.s. [Report on Research and Development Progress], 18 Feb. 1925, in Purtee, Appendix G.

Wooden fuselages apparently absorbed engine vibrations, but metal fuselages suffered fatigue failures. McCook Field engineers were developing rubber shock absorbers for engine mounts to extend their useful life. The Materiel Division contracted with Westinghouse Electric and Manufacturing Company for a study of aircraft structural and engine vibrations to supplement the work of the division's engineers. A.M. Jacobs, "Damping Out Engine Vibration," Air Corps News Letter, XI:3 (10 March 1927), 65-66.

76 "Two Years' Work at McCook Field," Aviation and Aircraft Journal, X:9 (28 Feb. 1921), 264. Eng. Div. Rpt. atch. to ltr, Maj. W.G. Kilner, Exec, AS, to Ch, Eng. Div, AS, n.s. [Report on Research and Development Progress], 18 Feb. 1925, in Purtee, Appendix G.

77 Eng. Div. Rpt. Nr. 238.

78 Lt. Cmdr. Bruce G. Leighton, USN, Journal of the Society of Automotive Engineers, (April 1924), as cited in Eng. Div. Rpt. Nr. 238.

79 Boyne, Wings, 12.

80 Brig. Gen. William E. Gillmore, Ch, Matl. Div, AAC, "The Development of Air-Cooled Engines," Airway Age, IX:7 (July 1928), 13-14. Charles L. Lawrance, "Air Cooled Engine Development," Aviation, XXIII:13 (26 Sep. 1927), 723-725. Termena, Peiffer, and Carlin, 38. Wilson, Autobiography, 33-36. Ltr, E.T. Jones, Ch, Power Plant Sect, Eng. Div, AS, to Ch, Eng. Div, AS, subj: Data on Power Plant Development, 13 Jan. 1925, in Purtee, Appendix F. Heron, Piston Engine, 21-26, 108. Taylor, 64-65. Byttebier, 69. Schalifer, 31, 124-125, 177-178.

81 Heron gave credit for his work directly to England and indirectly to an Italian designer, unknown to him, who influenced early English work on aluminum cylinders. S.D. Heron, "Origin of the Modern Air-Cooled Cylinder," Metal Progress (Jan. 1955), 137-140. S.D. Heron, "Air-Cooled Airplane Engines," U.S. Air Service, VII:4 (May 1922), 26-27. S.D. Heron, "Air-Cooled Engine Development," Mechanical Engineering, (Oct. 1925), 791-793. "The Power Plant Section: McCook Field, Dayton, Ohio," Slipstream, IV:8 (Aug. 1923), 34. Ltr, E.T. Jones, Ch, Power Plant Sect, Eng. Div, AS, to Ch. Engr, Eng. Div, AS, subj: Data on Power Plant Development, 13 Jan. 1925, in Purtee, Appendix F. Charles L. Lawrance, "Air Cooled Engine Development," Aviation, XXIII:13 (26 Sep. 1927), 723-725. Byttebier, 69.

Heron credited McCook Field's Materials Section foundry with producing "the first satisfactory air-cooled cylinder castings made in the United States," adding:

> Dix was responsible for making the first satisfactory 'Y' alloy air-cooled cylinder head castings. Maybe I should have said more and stated that the McCook Field foundry was responsible for making the first satisfactory air-cooled cylinder castings in this country, not only in 'Y' alloy.

Heron, Autobiography, 356. For an explanation of how the Engineering Division solved the casting of air-cooled cylinders, see Edgar H. Dix Jr., "Foundry Production of Air-Cooled Cylinders," Journal of the Society of Automotive Engineers, XII:1 (Jan. 1923), Appendix 2, 53-56.
McCook Field's foundry was the first in the U.S. to make castings of magnesium alloys. The Engineering Division took pride in its support of magnesium development, pointing out that work started as early as

1920 and that the first magnesium parts used by the Curtiss and Packard companies were cast in the division's foundry. By 1925, work on casting magnesium alloys progressed to the point where the foundry was turning out alloys in the form of instrument cases, filler caps, rocker boxes, and other aircraft parts. Stepping up of the process was expected to produce large castings like crankcases and cylinder blocks.

Eng. Div. Rpt. Nr. 238. Hobert L. Wilson, "Army Engineering Developments of Interest to Commercial Aviation," Slipstream, VI:8 (Aug. 1925), 5. McCook Field brochure.

Brig. Gen. William E. Gillmore, Ch, Matl. Div., AAC, "The Development of Air-Cooled Engines," Airway Age, IX:7 (July 1928), 13-14. For discussion of air-cooled cylinders as they developed during the McCook Field era, see C. Fayette Taylor, "The Design of Air-Cooled Cylinders," Slipstream, VI:6 & 7 (June & July 1925), 13-15, 22; & 20-24; also published in Aviation, XVIII:23 & 24 (8 & 15 June 1925), 634-636 & 664-667. Heron, Piston Engine, 21-25, 108-109.

[82] Schlaifer, 162-165, 176. "The Lawrance 3-Cylinder Airplane Engine," Aviation, VI:3 (1 March 1919), 141. "Lawrance L-2 Air-Cooled Motor," Aircraft Journal, VI:14 (3 April 1920), 7-8. "The Gas Bag," Slipstream, III:15 (15 Sep. 1921), 6. George S. Wheat, "The Air-Cooled Engine," Aero Digest, V:6 (Dec. 1924), 367-370. "Navy Rejoices at Great Success of Air-Cooled Engines Which It Sponsored," U.S. Air Services, XII:8 (Aug. 1927), 48, 50. Charles L. Lawrance, "Air Cooled Engine Development," Aviation, XXIII:13 (26 Sep. 1927), 723-725.

[83] Maj. A.H. Hobley, Asst. Ch, Eng. Div, AS, "Achievements of the Engineering Division for the Calendar Year 1922," Slipstream, IV:2 (Feb. 1923), 6. "Two Years' Work at McCook Field," Aviation and Aircraft Journal, X:9 (28 Feb. 1921), 264. Carr article. McCook Field brochure. George Wheat, "The Air-Cooled Engine," Aero Digest, V:6 (Dec. 1924), 367-370. "The Commercial Use of Wright Whirlwind 200 h.p. Air Cooled Engines," Aviation, XX:3 (18 Jan. 1926), 79-80. Lt. Cmdr. E.E. Wilson, USN, "Aircraft Engine Progress," U.S. Air Services, X:2 (Feb. 1925), 15-19. Cmdr. E.E. Wilson, USN, "Air-Cooled and Water-Cooled Engines," U.S. Air Services, XI:6 (June 1926), 35-38. Cmdr. E.E. Wilson, USN, "Navy Air-Cooled Engine Development," Aviation, XXI:2 (12 July 1926), 59-61. "Great Improvements Shown in Engines," U.S. Air Services, XII:1 (Jan. 1927), 44. Cmdr. E.E. Wilson, USN, "The Recent Air Records," Aviation, XXII:22 (30 May 1927), 1129-1130. "History and Development of Wright Whirlwind," Slipstream, VIII:8 (Aug. 1927), 10. "Navy Rejoices at Great Success of Air-Cooled Engines Which It Sponsored," U.S. Air Services, XII:8 (Aug. 1927), 48, 50. Charles L. Lawrance, "Air Cooled Engine Development," Aviation, XXIII:13 (26 Sep. 1927), 723-725. Cmdr. E.E. Wilson, USN, "American Aircooled Aircraft Engines," Aero Digest, XI:3 (Sep. 1927), 272, 358-359. Brig. Gen. William E. Gillmore, Ch, Matl. Div., AAC, "The Development of Air-Cooled Engines," Airway Age, IX:7 (July 1928), 13-14. Capt. Burdette S. Wright, "Progress of the Air Corps Under Administration of Maj. Gen. Mason M. Patrick," Slipstream, VIII:12 (Dec. 1927), 16. Termena, Peiffer, and Carlin, 32, 34. Wilson, Air Power, 63-64. Heron, Piston Engine, 30. Nayler, 149. Byttebier, 76-77. Taylor, 41, 43, 53. Schlaifer, 28, 62-63, 165-176.

Citing figures from U.S. air mail flying, the Wright Aeronautical Corporation indicated that radiator-related troubles caused 30 percent of the aircraft problems, ignition 20 percent, carburetion 11 percent, and lubrication 8 percent -- a total of 78 percent of forced landings in single-engine aircraft. The answer urged by Wright Aeronautical: use of three air-cooled engines on passenger aircraft. C.G. Peterson, "Three-Engine Planes for Air Transport," Aviation, XXI:9 (30 Aug. 1926), 354-357.
By mid-1927, over 1,000 Whirlwinds were delivered for use by the Army and Navy and mmercial firms, including more than 270 for the latter. "Wright Whirlwinds Have Played Important Parts in Making Aeronautical History," Aviation, XXII:26 (27 June 1927), 1432-1433.

Detailed description of the manufacturing process used in producing the Whirlwind, from foundry to flight, was provided by Lee M. Beatty, "Wright Whirlwind Engine Production," Aviation, XXIII:13 (26 Sep. 1927), 727-732, 767.

In 1926, Lieutenant Commander Richard E. Byrd did a pioneer exploration of the Arctic with three Loening amphibians. Commander Byrd flew over the North Pole on 9 May 1926 in a Fokker monoplane equipped with three Wright Whirlwind engines. His pilot was Floyd Bennett. The aircraft was also equipped with a Curtiss-Reed metal propeller. Loening, Takeoff, 161, 192. "Byrd First to Reach North Pole Via Air," Slipstream, VII:6 (June 1926), 7.

84   Wilson, Autobiography, 12-14, 17-19, 29-32, 36, 38-39, 52-53. See also, Lt. Cmdr. Bruce G. Leighton, USN, "The Airplane Power Plant and Safe Flying," Slipstream, VIII: 11 (Nov. 1927), 21-23, 30. K.J. Boedecker, "The Economy of Air Cooling," Aviation, XVIII:18 (4 May 1925), 492-493. A.H.R. Fedden, "Radial Air-Cooled Aero Engines," Slipstream, VII:2 (Feb. 1926), 16-17; and VII:3 (March 1926), 18-19. Taylor, 53.

      Although committed to air cooling, the Navy signed large orders in 1926 for Packard 1A-1500 500-horsepower and 2A-2500 800-horsepower engines and for Curtiss D-12 engines to be used on carrier aircraft for the Saratoga and Lexington. "Navy Places Large Aircraft Orders," Aviation, XX:13 (29 March 1926), 450. See also, Lt. Cmdr. F.W. Wead, USN, "Air-Cooled Fighters or Water-Cooled?" Aviation, XXII:12 (21 March 1927), 565-567.

85   "McCook Field, Dayton, Ohio, Dec. 1," Air Service News Letter, V:45 (27 Dec. 1921, 11. Rpt, C.F. Taylor, Power Plant Sect, Eng. Div, AS, "Aircraft Development Since the Armistice - Engines," 7 June 1922. "The Power Plant Section: McCook Field, Dayton, Ohio," Slipstream, IV:8 (Aug. 1923), 34. Maj. A.H. Hobley, Asst. Ch, Eng. Div, AS, "Achievements of the Engineering Division for the Calendar Year 1922," Slipstream, IV:2 (Feb. 1923), 6. Rpt, Digest of the Report of Achievements of the Engineering Division, Air Service, for Period 1920-1922, n.d. [c. 1922]. Charles L. Lawrance, "Air Cooled Engine Development," Aviation, XXIII:13 (26 Sep. 1927), 723-725. Schlaifer, 173, 176-179. Heron, Autobiography, 85-105, 112-117, 157-167.

86   Eng. Div. Rpt. atch. to ltr, Maj. W.G. Kilner, Exec, AS, to Ch, Eng. Div, AS, n.s. [Report on Research and Development Progress], 18 Feb. 1925, in Purtee, Appendix G. Schlaifer, 29.

87   Eng. Div. Rpt. atch. to ltr, Maj. W.G. Kilner, Exec, AS, to Ch, Eng. Div, AS, n.s. [Report on Research and Development Progress], 18 Feb. 1925, in Purtee, Appendix G. "The Mark VI Bristol Jupiter Engine," Aero Digest, XI:3 (Sep. 1927), 290, 292.

88   Eng. Div. Rpt. atch. to ltr, Maj. W.G. Kilner, Exec, AS, to Ch, Eng. Div, AS, n.s. [Report on Research and Development Progress], 18 Feb. 1925, in Purtee, Appendix G. Maj. A.H. Hobley, Asst. Ch, Eng. Div, AS, "Achievements of the Engineering Division for the Calendar Year 1922," Slipstream, IV:2 (Feb. 1923), 6.

89   Schlaifer, 18-20, 23-24, 62-65, 179-181, and 184.

90   Rpt, Maj. Leslie MacDill, Ch, Exp. Eng. Sect, Matl. Div, AAC, subj: Status of Experimental and Service Test Projects, Engineering Program - Fiscal Year 1927, As of 1 January 1927, 10 Jan. 1927. Brig. Gen. William E. Gillmore, Ch, Matl. Div, AAC, "Review of Year's Developments in the Army Air Corps," Slipstream, VIII:3 (March 1927), 12. Rpt, Brig. Gen. William E. Gillmore, Ch, Matl. Div, AAC, subj: Second Annual Report of the Chief, Materiel Division, Air Corps, Fiscal Year 1928 [1 July 1927-30 June 1928]. Ltr, E.T. Jones, Ch, Power Plant Sect, Eng. Div, AS, to Ch, Eng. Div, AS, subj: Data on Power Plant Development, 13 Jan. 1925, in Purtee, Appendix F. Eng. Div. Rpt. atch. to ltr, Maj. W.G. Kilner, Exec, AS, to Ch, Eng. Div, AS, n.s. [Report on Research and Development Progress], 18 Feb. 1925, in Purtee, Appendix G. Maj. John F. Curry, Ch, Eng. Div, AS, "McCook Field Review," Aviation, XX:2 (11 Jan. 1926), 48. Brig. Gen. William E. Gillmore, Ch, Matl. Div, AAC, "The Development of Air-Cooled

Engines," Airway Age, IX:7 (July 1928), 13-14. Wilson, Autobiography, 34-36, 50-52. Heron, Piston Engine, 27-28, 39. Byttebier, 104. Heron, Autobiography, 79-85, 103, 105, 112-117, 123-124.

91   Hobert L. Wilson, "Air-Cooling the Liberty Engine," Slipstream, VI:12 (Dec. 1925), 18. Ltr, E.T. Jones, Ch, Power Plant Sect, Eng. Div, AS, to Ch, Eng. Div, AS, subj: Data on Power Plant Development, 13 Jan. 1925, in Purtee, Appendix F. N.H. Gilman, Ch. Engr, Allison Eng. Co, "The Air Cooled Liberty Engine," Aviation, XXIII:25 (19 Dec. 1927), 1468-1470. Heron, Autobiography, 101, 105-112.

92   "The Fairchild-Caminez Engine," Aviation, XX:21 (24 May 1926), 788-791. "Cam Engine Passes Fifty Hour Test," Aviation, XXIII:1 (4 July 1927), 20-21. "The New Cam Engine," Slipstream, VIII:8 (Aug. 1927), 17-18, 20. "Development of the Cam Type Aviation Engine," Aero Digest, XI:2 (Aug. 1927), 158, 160. "Fairchild Caminez Engine Produces Interesting Figures," U.S. Air Services, XII:12 (Dec. 1927), 36-37. Harold Caminez, "Technical Aspects of the Fairchild Caminez Engine," Airway Age, IX:7 (July 1928), 19-21. Taylor, 57.

93   Brig. Gen. William E. Gillmore, Ch, Matl. Div, AAC, "The Development of Air-Cooled Engines," Airway Age, IX:7 (July 1928), 13-14. Rpt, Brig. Gen. William E. Gillmore, Ch, Matl. Div, AAC, subj: Second Annual Report of the Chief, Materiel Division, Air Corps, Fiscal Year 1928 [1 July 1927-30 June 1928]. Aircraft Year Book 1927, 80, 211, 213, 230-231. "The Pratt & Whitney Wasp Engine," Aviation, XX:22 (31 May 1926), 827-828. "Pratt & Whitney Anniversary," Aviation, XXI:6 (9 Aug. 1926), 246, 248-249. F.B. Rentschler, "Development of the 'Wasp,'" Slipstream, VII:11 (Nov. 1926), 13-14. "Great Improvements Shown in Engines," U.S. Air Services, XII:1 (Jan. 1927), 44, 46. "Pratt & Whitney Aircraft Corporation Starts Big Production of 'Wasp,'" Slipstream, VIII:4 (April 1927), 17. "Pratt & Whitney Hornet Engine Successfully Passes Navy Fifty Hour Type Test," Aviation, XXII:18 (2 May 1927), 897-899. Cmdr. E.E. Wilson, USN, "New Navy Aircraft," U.S. Air Services, XII:5 (May 1927), 19-21. "Navy Rejoices at Great Success of Air-Cooled Engines Which It Sponsored," U.S. Air Services, XII:8 (Aug. 1927), 48, 50. Charles L. Lawrance, "Air Cooled Engine Development," Aviation, XXIII:13 (26 Sep. 1927), 723-725. "The 425 h.p. Pratt and Whitney Wasp Engine," Aero Digest, XI:3 (Sep. 1927), 298, 300. C.W. Deeds, "Modern Air Cooled Motors in Commercial Service," Slipstream, VIII:10 (Oct. 1927), 36-37. "Personals," Slipstream, I:9 (29 Feb. 1920), 11. Termena, Peiffer, and Carlin, 32, 34. Loening, Takeoff, 188-190. Wilson, Air Power, 64, 70. Wilson, Autobiography, 48-56, 72-77-80, 93-94. Taylor, 45-46. Schlaifer, 18-19, 23-24, 65-67, 77, 182-196. Heron, Piston Engine, 39-40. Heron, Autobiography, 142-149.

Responding to gibes that his fame was dim compared to Lindbergh's heroic repute after flying from New York to Paris with a Whirlwind engine, Lawrance asked: "Who remembers the name of Paul Revere's horse?" Loening, Takeoff, 134.

Progress by the U.S. in engine technology, including water- and air-cooled engines, valves, pistons, cylinders, gearing, and supercharging was described before the Royal Aeronautical Society in England by Charles L. Lawrance, "Modern American Aircraft Engine Development," Aviation, XX:11 & 12 (15 & 22 March 1926), 363(?)-367 & 411-415.

A brief history of air-cooled radial engines and their potential for commercial application by the mid-1920s was in Robert W.A. Brewer, "The Air-Cooled Radial Engine," Aviation, XX:23 (7 June 1926), 872-874. He also addressed problems related to installing such engines on aircraft in "Adaptation of the Radial Air-Cooled Engine," Aviation, XX:25 (21 June 1926), 942-944.

Compared to the Liberty engine, the air-cooled Cyclone represented a major advance. The Liberty engine's installed weight was 3.2 pounds per horsepower versus the Cyclone's 2 pounds per horsepower. Other advantages of the Cyclone engine, originally begun by the Lawrance Aero Engine Corporation but developed after that company consolidated with Wright Aeronautical Corporation, were described by C.G. Peterson, "Wright 'Cyclone' Engine," Aero Digest, VI:6 (June 1925), 304, 336. See also, "Wright

'Cyclone" an Improvement," and "Recent Developments in Aero Engines," Slipstream, VI:5 (May 1925), 17, and 19; and "Wright Cyclone Engine Flight Tested," Aviation, XIX:8 (24 Aug. 1925), 211.

By the time of World War II, Pratt & Whitney radial engines matured into the 18-cylinder R-2800 series and the Wasp Major R-4360 of 1944 with 28-cylinders in four 7-cylinder banks which generated 3,500-horsepower. This engine marked the end of the radial air-cooled era -- jets were on the way. By 1945, with WWII production included, P&W delivered almost 500,000 air-cooled radial engines for military and commercial use. Loening, Takeoff, 190-191. Wilson, Autobiography, 285. Heron, Piston Engine, 30-31. Nayler, 143, 149-151. Schlaifer, 79.

In 1929, Wright merged with Curtiss, becoming the Curtiss-Wright Corporation, and the new firm continued to evolve larger engines in the Cyclone series with multibanked cylinder configurations and with power equal to and sometimes larger than the Wasp series. But the Wright Whirlwind descendants never attained the excellence, reliability, and quantity-production-on-time of the Pratt & Whitney engines. Loening, Takeoff, 191. Nayler, 150-151.

94 Hobert L. Wilson, "Army Engineering Developments of Interest to Commercial Aviation," Slipstream, VI:8 (Aug. 1925), 5.

95 Further reductions in cooling drag were achieved by increasing the cooling-fin area, thereby reducing the air velocity required for cooling. These developments put the air-cooled radial virtually on a par with the water-cooled engines with regard to cooling drag until the application of high-temperature liquid cooling with glycol-water mixtures. "The Power Plant Section: McCook Field, Dayton, Ohio," Slipstream, IV:8 (Aug. 1923), 34. Schlaifer, 171. Byttebier, 82. Taylor, 53, 55.

96 Cmdr. E.E. Wilson, USN, "Power Plant Aspects of the National Air Races," Aviation, XXI:13 (27 Sep. 1926), 543-544. Byttebier, 81-82.

97 Cammen article, 442. Eng. Div. Rpt. Nr. 238. Rpt, David Gregg, Supercharger Br, Power Plant Sect, Eng. Div, AS, "A Comparison of the Performances of Supercharged and Unsupercharged Airplanes, 3 Aug. 1922. For technical discussion, see Sanford Alexander Moss, Superchargers for Aviation (New York: National Aeronautics Council, Inc., 1942.) Nayler, 151-152. Taylor, 68. Heron, Piston Engine, 39-40, 44. Heron, Autobiography, 125-126.

98 Eng. Div. Rpt. atch. to ltr, Maj. W.G. Kilner, Exec, AS, to Ch, Eng. Div, AS, n.s. [Report on Research and Development Progress], 18 Feb. 1925, in Purtee, Appendix G.

99 "The Power Plant Section: McCook Field, Dayton, Ohio," Slipstream (Aug. 1923), 35. "Two Years' Work at McCook Field," Aviation and Aircraft Journal, X:9 (28 Feb. 1921), 263-265. Rpt, Digest of the Report of Achievements of the Engineering Division, Air Service, for Period 1920-1922, n.d. [c. 1922]. Carr article. Nayler, 145, 151.

100 Contemporary progress on supercharger development was provided in several articles. Maj. George E.A. Hallett, Ch, Power Plant Sect, Eng. Div, AS, "Superchargers and Supercharging Engines," General Electric Review, XXIII:6 (June 1920), 468-473, also published in Aviation, VII:12 (15 Jan. 1920), 533-536; Hallett presented the paper before the Society of Automotive Engineers. Dr. Sanford A. Moss, "The General Electric Turbo-Supercharger for Airplanes," General Electric Review, XXIII:6 (June 1920), 476-485, also published in Aviation, VIII:4 (15 March 1920), 146-151. Rpt, C.F. Taylor, Power Plant Sect, Eng. Div, AS, "Aircraft Development Since the Armistice - Engines," 7 June 1922. David Gregg, Supercharger Research Dept, McCook Field, "The Story of the Supercharger," Slipstream, V:3 (March 1924), 17-19, 24. David Gregg, "Superchargers," Aviation, XIX:4 (27 July 1925), 90-93. Lt. Cmdr. F.W. Wead, USN, "Next Step, Superchargers?" Aviation, XXIII: 12 (19 Sep. 1927), 671-672 ff. An assessment was provided by A.L. Berger and Opie Chenoweth, Exp. Eng. Sect, Matl. Div, AAC, "The Turbo Super-

charger," a paper prepared for presentation at the 20th National Aeronautic Meeting of the Society of Automotive Engineers at Cleveland, 1-3 Sep. 1931. Heron, Piston Engine, 44. Taylor, 69.

Taylor claimed that Jones used the side-type turbo-supercharger at his suggestion; however, Heron rejected Taylor's claim and indicated that Hallett also rejected Taylor's claim. Taylor, 69, 71. Heron, Autobiography, 125-127.

101   Rpt, Maj. Leslie MacDill, Ch, Exp. Eng. Sect, Matl. Div, AAC, subj: Status of Experimental and Service Test Projects, Engineering Program - Fiscal Year 1927, As of 1 January 1927, 10 Jan. 1927. Brig. Gen. William E. Gillmore, Ch, Matl. Div, AAC, "Review of Year's Developments in the Army Air Corps," Slipstream, VIII:3 (March 1927), 13-14. Rpts, Brig. Gen. William E. Gillmore, Ch, Matl. Div, AAC, subj: First and Second Annual Reports of the Chief, Materiel Division, Air Corps, Fiscal Year 1927 [1 July 1926-30 June 1927] & Fiscal Year 1928 [1 July 1927-30 June 1928]. David Gregg, "Commercial Superchargers," Aero Digest, X:1 (Jan. 1927), 20, 70-71. Ltr, E.T. Jones, Ch, Power Plant Sect, Eng. Div, AS, to Ch. Engr, Eng. Div, AS, subj: Data on Power Plant Development, 13 Jan. 1925, in Purtee, Appendix F. Eng. Div. Rpt. atch. to ltr, Maj. W.G. Kilner, Exec, AS, to Ch, Eng. Div, AS, n.s. [Report on Research and Development Progress], 18 Feb. 1925, in Purtee, Appendix G. Aircraft Year Book 1927, 215. Hobert L. Wilson, "Army Engineering Developments of Interest to Commercial Aviation," Slipstream, VI:8 (Aug. 1925), 8. Lt. Cmdr. F.W. Wead, USN, "Next Step, Superchargers?" Aviation, XXIII:12 (19 Sep. 1927), 671-672 ff. Heron, Piston Engine, 37-41, 47. Nayler, 152.

102   Eng. Div. Rpt. atch. to ltr, Maj. W.G. Kilner, Exec, AS, to Ch, Eng. Div, AS, n.s. [Report on Research and Development Progress], 18 Feb. 1925, in Purtee, Appendix G. E.T. Jones, "Achievements of American Aircraft-Engine Industry," Mechanical Engineering (Oct. 1925), 790-791. Slipstream (Aug. 1923), 35. McCook Field brochure. Heron, Piston Engine, 44-46. Heron, Autobiography, 125-126, 130-131.

103   Eng. Div. Rpt. atch. to ltr, Maj. W.G. Kilner, Exec, AS, to Ch, Eng. Div, AS, n.s. [Report on Research and Development Progress], 18 Feb. 1925, in Purtee, Appendix G.

104   Schlaifer, 120.

Schlaifer, pages 102-103, faults GE's lack of competition in supercharger development with causing American superchargers to be "seriously inferior" to Britain's from their first appearance on production engines in 1927. GE's monopoly kept engine companies from realizing greater efficiency and performance in their products. Progress in the U.S. accelerated in 1937 when Wright Aeronautical began designing its own superchargers.

105   E.W. Dean and Clarence Netzen, "An Investigation of Airplane Fuels," Aviation, VII:12 (15 Jan. 1920), 537-538.

106   Carr article. Rpt, C.F. Taylor, Power Plant Sect, Eng. Div, AS, "Aircraft Development Since the Armistice - Engines," 7 June 1922. Ltr., E.T. Jones, Ch, Power Plant Sect, Eng. Div, AS, to Ch. Engr, Eng. Div, AS, subj: Data on Power Plant Development, 13 Jan. 1925, in Purtee, Appendix F. Hobert L. Wilson, "Army Engineering Developments of Interest to Commercial Aviation," Slipstream, VI:8 (Aug. 1925), 6. E.T. Jones, Ch. Engr, Wright Aeronautical Corp, "Fuel for the Wright 'Whirlwind,'" Aviation, XXIII:20 (14 Nov. 1927), 1170-1172.

For discussion of developments with fuels and fuel systems, see the chapters in Heron, Piston Engines. See also, Nayler, 156. Taylor, 65-66. Schlaifer, 29, 31-32. Heron, Autobiography, 135, 191.

107   "Engineering Branch of Shop Engineering Section," Slipstream, I:3 (15 Sep. 1919), 32. "The Airplane Section," Slipstream, II:5 (1 Sep. 1920), 2. Taylor, 77. See also, Edward T. Jones, Robert Insley, Frank

W. Caldwell, and Robert F. Kohr, <u>Aircraft Power Plants</u> (New York: Ronald Press Co, 1926), Part II: "Propellers."

For discussion of propeller development from WWI through WWII, especially at McCook and Wright Fields, see Dickey intvw. Some of McCook Field's contributions to metal propeller technology were also summarized by the Engineering Division's Ernest G. McCauley, "Metal Propeller Development," <u>Aviation</u>, XXII:22 (30 May 1927), 1127-1130, 1144.

108   "Wood Shops," Slipstream, I:7 (15 Dec. 1919), 44.

109   Lt. Col. Whiston A. Bristow, "The Design and Construction of Metal Propellers," <u>Slipstream</u>, VII:7 & 8 (July & Aug. 1926), 22-23, 31; and 16-19. Cammen article, 442. H.S. Hele-Shaw and T.E. Beacham, "The Variable Pitch Propeller," <u>Airway Age</u>, IX:8 (Aug. 1928), 35-39. Nayler, 157.

110   "Reversible Propeller," <u>Slipstream</u>, I:6 (1 Nov. 1919), 6.

111   "Micarta Propellers on Liberty Motors," <u>Slipstream</u>, I:7 (15 Dec. 1919), 21, 31. "Propellers: A Description of the Propeller Branch, McCook Field." <u>Slipstream</u> IV:8 (Aug. 1923), 9-10. "Service Planes to Have Metal Propellers," <u>Slipstream</u>, V:3 (March 1924), 36. See also, Dickey intvw.

112   "Propellers: A Description of the Propeller Branch, McCook Field," <u>Slipstream</u> IV:8 (Aug. 1923), 9-10.

113   "Propellers: A Description of the Propeller Branch, McCook Field," <u>Slipstream</u> IV:8 (Aug. 1923), 9-10.

114   "Propellers: A Description of the Propeller Branch, McCook Field," <u>Slipstream</u> IV:8 (Aug. 1923), 9-10. McCook Field brochure.

115   Taylor, 77.

116   "Propellers: A Description of the Propeller Branch, McCook Field," <u>Slipstream</u> IV:8 (Aug. 1923), 9-10. Eng. Div. Rpt. atch. to ltr, Maj. W.G. Kilner, Exec, AS, to Ch, Eng. Div, AS, n.s. [Report on Research and Development Progress], 18 Feb. 1925, <u>in</u> Purtee, Appendix G.

117   "Propellers: A Description of the Propeller Branch, McCook Field," <u>Slipstream</u> IV:8 (Aug. 1923), 9-10. "Mammoth Propeller Turned Out at McCook," <u>Slipstream</u>, IV:4 (April 1923), 24. "Two Years' Work at McCook Field," <u>Aviation and Aircraft Journal</u>, X:9 (28 Feb. 1921), 263-265. Rpt, Digest of the Report of Achievements of the Engineering Division, Air Service, for Period 1920-1922, n.d. [c. 1922]. R.I. Markey, "The Spotlight on McCook Field," <u>The Ohio State Engineer</u> (Nov. 1920), 10.

118   "Propellers: A Description of the Propeller Branch, McCook Field," <u>Slipstream</u> IV:8 (Aug. 1923), 9-10. Maj. A.H. Hobley, Asst. Ch, Eng. Div, AS, "Achievements of the Engineering Division for the Calendar Year 1922," <u>Slipstream</u>, IV:1 (Jan. 1923), 23. Eng. Div. Rpt. atch. to ltr, Maj. W.G. Kilner, Exec, AS, to Ch, Eng. Div, AS, n.s. [Report on Research and Development Progress], 18 Feb. 1925, <u>in</u> Purtee, Appendix G. "Two Years' Work at McCook Field," <u>Aviation and Aircraft Journal</u>, X:9 (28 Feb. 1921), 263-265. Dickey intvw. Ernest G. McCauley, "Metal Propeller Development," <u>Aviation</u>, XXII:22 (30 May 1927), 1127-1130, 1144.

119   "Propellers: A Description of the Propeller Branch, McCook Field," <u>Slipstream</u> IV:8 (Aug. 1923), 9-10.

120   "Propellers: A Description of the Propeller Branch, McCook Field," <u>Slipstream</u> IV:8 (Aug. 1923), 9-10. Eng. Div. Rpt. atch. to ltr, Maj. W.G. Kilner, Exec, AS, to Ch, Eng. Div, AS, n.s. [Report on Research and Development Progress], 18 Feb. 1925, <u>in</u> Purtee, Appendix G.

121 Lt. Col. Whiston A. Bristow, "The Design and Construction of Metal Propellers," Slipstream, VII:7 & 8 (July & Aug. 1926), 22-23, 31; and 16-19. Lt. S.H. Wooster, USN, "The Case for the Metal Propeller," U.S. Air Services, X:5 (May 1925), 24-27. Nayler, 151, 157. Dickey intvw.

122 Brig. Gen. William E. Gillmore, Ch, Matl. Div, AAC, "Review of Year's Developments in the Army Air Corps," Slipstream, VIII:3 (March 1927), 10. Eng. Div. Rpt. atch. to ltr, Maj. W.G. Kilner, Exec, AS, to Ch, Eng. Div, AS, n.s. [Report on Research and Development Progress], 18 Feb. 1925, in Purtee, Appendix G. "Development of Curtiss-Reed Duralumin Propeller," Slipstream, VI:1 (Jan. 1925), 27. "First Magnesium Propeller Meets Test Well," U.S. Air Services, X:4 (April 1925), 47. Lt. S.H. Wooster, USN, "The Case for the Metal Propeller," U.S. Air Services, X:5 (May 1925), 24-27. T.P. Wright, "The Durability of Metal Propellers," Aviation, XXI:17 (25 Oct. 1926), 706. R.B.C. Noorduyn & T.P. Wright, "Metal v. Wood Propellers," Aviation, XXI:22 (29 Nov. 1926), 913-914. S.A. Reed & Arthur D. Sallee, "Further Discussion on Metal Propellers," Aviation, XXI:26 (27 Dec. 1926), 1082-1083. "Review of Curtiss Aeroplane & Motor Co., Inc.," Slipstream, VIII:4 (April 1927), 20, 22, 24, 27. Aircraft Year Book 1927, 213, 215, 232-233. S. Albert Reed, "Technical Development of the Reed Metal Propeller," Airway Age, IX:10 (Oct. 1928), 32-33. Wilson, Autobiography, 77-78. Heron, Piston Engines, 55-56. Nayler, 157. Taylor, 77-78.

123 Lt. Col. Whiston A. Bristow, "The Design and Construction of Metal Propellers," Slipstream, VII:7 & 8 (July & Aug. 1926), 22-23, 31; and 16-19. Dickey intvw.

124 Lt. Col. Whiston A. Bristow, "The Design and Construction of Metal Propellers," Slipstream, VII:7 & 8 (July & Aug. 1926), 22-23, 31; and 16-19.

125 Lt. Col. Whiston A. Bristow, "The Design and Construction of Metal Propellers," Slipstream, VII:7 & 8 (July & Aug. 1926), 22-23, 31; and 16-19. Dickey intvw.

126 Lt. Col. Whiston A. Bristow, "The Design and Construction of Metal Propellers," Slipstream, VII:7 & 8 (July & Aug. 1926), 22-23, 31; and 16-19. "The Leitner-Watts Metal Propeller," Aviation, XIX:9 (31 Aug. 1925), 244-245.

127 Byttebier, 60.

128 Byttebier, 60. F.H. Russell, "Development of the Curtiss-Reed Propeller," Slipstream, VI:1 (Jan. 1925), 27-28. Sylvanus Albert Reed, "Technical Development of the Reed Metal Propeller," Airway Age, IX:10 (Oct. 1928), 32-33.

129 F.H. Russell, "Development of the Curtiss-Reed Propeller," Slipstream, VI:1 (Jan. 1925), 27-28. Lt. Col. Whiston A. Bristow, "The Design and Construction of Metal Propellers," Slipstream, VII:7 & 8 (July & Aug. 1926), 22-23, 31; and 16-19. Dickey intvw.

130 F.H. Russell, "Development of the Curtiss-Reed Propeller," Slipstream, VI:1 (Jan. 1925), 27-28. Lt. Col. Whiston A. Bristow, "The Design and Construction of Metal Propellers," Slipstream, VII:7 & 8 (July & Aug. 1926), 22-23, 31; and 16-19.

131 F.H. Russell, "Development of the Curtiss-Reed Propeller," Slipstream, VI:1 (Jan. 1925), 27-28. Byttebier, 62-63.

132 F.H. Russell, "Development of the Curtiss-Reed Propeller," Slipstream, VI:1 (Jan. 1925), 27-28. Lt. Col. Whiston A. Bristow, "The Design and Construction of Metal Propellers," Slipstream, VII:7 & 8 (July & Aug. 1926), 22-23, 31; and 16-19.

133 "First Magnesium Propeller a Success," Slipstream, VI:3 (March 1925), 25. "First Magnesium Propeller Successfully Tested," Aviation, XVIII:9 (2 March 1925), 248. "First Magnesium Propeller Meets Test Well," U.S. Air Services, X:4 (April 1925), 47. Eng. Div. Rpt. atch. to ltr, Maj. W.G. Kilner, Exec, AS, to Ch, Eng. Div, AS, n.s. [Report on Research and Development Progress], 18 Feb. 1925, in Purtee, Appendix G. Dickey intvw.

134 "Dr. S. Albert Reed Accepts the Collier Trophy From N.A.A.," U.S. Air Services, XI:5 (May 1926), 27.

135 "Aero Development from Design and Experiment in Ohio," Slipstream, VIII:9 (Sep. 1927), 37. Compare Col. Thurman H. Bane, Ch, Eng. Div, AS, "Recent Advances in Aviation," Society of Automotive Engineers Transactions, XV (1920), Part II, 63-86; and F. Trubee Davison, Asst. Sec/War, "The Place Wright Field Fills," Slipstream, VIII:10 (Oct. 1927), 21-22.

136 Boyne, Airpower, 10; and Wings, 8. Gross paper.

Grover Loening, who described aviation in the 1920s variously as a "golden era" and as a "mess" (see his Takeoff, 153 and 167) hailed 1927 as a pivotal year. "Loening Reviews Breathless Year in Aviation," U.S. Air Services, XII:12 (Dec. 1927), 24. See also, Adrian Van Muffling, "A Victorious Year of American Aeronautics -- 1927," Aero Digest, XI:6 (Dec. 1927), 642-643, 762-763.

137 Claussen, 20-21. "Army and Navy Procurement," Aircraft Year Book 1934, 112-113.

138 Aviation, VII:8 (15 Nov. 1919), 8.

139 "McCook Field and the Airplane Industry," Aviation, VII:12 (15 Jan. 1920), 525.

140 "McCook Field and the Airplane Industry," Aviation, VII:12 (15 Jan. 1920), 525.

141 Claussen, 21-22. Goldberg, 33. "Two Years' Work at McCook Field," Aviation and Aircraft Journal, X:9 (28 Feb. 1921), 263-265. Rpt, Digest of the Report of Achievements of the Engineering Division, Air Service, for Period 1920-1922, n.d. [c. 1922]. Carr article. Worman article, 14.

Claussen's list through 1922 included 14 designs and 27 aircraft actually built at McCook Field, including 12 bombers (USA-D-9, -9A, and -9AB; XB-1A); 2 attack (GA-1); 8 pursuit (PW-1, -1A; TP-1; TW-1; USA-O1; VCP-1; 4 observation (CO-1; CO-2); and 1 racer (R-1.) But Boyne, Airpower, 10, identified a total of 34 aircraft of 23 different designs turned out by the Engineering Division, and pointed out that still others were produced by manufacturers according to the division's designs. Wrote Boyne of the Engineering Division's models:

These ranged from tiny gliders and trainers to bombers and pursuits. Some were merely adaptations of well known foreign types; some were solid efforts of inspired engineers; some were wild forays into the improbable, and a few formed the cutting edge of the ever advancing state of the art.

142 Maj. Thurman H. Bane, Ch, Eng. Div, AS, "Notes from Speech by Major Bane," Slipstream, IV:1 (Jan. 1923), 42, 44. For more on Bane's efforts to strengthen commercial aviation through military contracting, see Maurer article, 26-27. See also, the extended discussion of "Army and Navy Procurement," Aircraft Year Book 1934, 109-126.

Efforts in 1919 by Colonel Bane to acquire 50 Martin MB-1 bombers for experimental purposes were nullified by Major General Charles T. Menoher who pointed out that British Handley-Page bombers were still in active inventory.

[143]   Claussen, 22-23.

[144]   Claussen, 23.

[145]   "Standard Engines of the Air Service," <u>Aviation</u>, XII:17 (24 April 1922), 480.

[146]   Purtee, 154-155. Quoted from Patrick, 100-105. See, for example, "Manufacturers' Bids for Construction of Martin Bombers," <u>Aviation</u>, XII:7 (13 Feb. 1922), 198. "Assistant Secretary of War Visits Dayton Fields," <u>Air Service News Letter</u>, VII:23 (17 Dec. 1923), 10-11. Capt. Burdette S. Wright, "Progress of the Air Corps Under Administration of Maj. Gen. Mason M. Patrick," <u>Slipstream</u>, VIII:12 (Dec. 1927), 22. Boyne, <u>Wings</u>, 18. Gross paper. Loening, <u>Takeoff</u>, 128-131.

A general description of the process for aircraft development was provided by an Ohio State University engineering student who was a summer employee at McCook Field: R.I. Markey, "The Spotlight on McCook Field," <u>The Ohio State Engineer</u> (Nov. 1920), 9-11. See also, "Experimental Types Ordered by Army," <u>Aviation</u>, IX:15 (27 Dec. 1920), 491-492. For an extended discussion, see "Army and Navy Procurement," <u>Aircraft Year Book 1934</u>, 114.

Insight into Major General Charles T. Menoher's procurement policy as chief of Air Service is contained in Purtee's Appendix E. Menoher expressed preference for purchasing materiel from the original designer. However, if the designer or developer did not charge a "reasonable" price, he directed the Air Service to secure bids or equipment from others.

Practical application of General Patrick's new acquisition policy was reflected in an invitation for a two-seat observation aircraft sent to "qualified manufacturers to submit complete airplanes of their own design and constructed in accordance with the specification and other requirements of the Air Service for this type for consideration and test." For details of the request, see "Air Service Asks for New Observation Planes," <u>Aviation</u>, XVI:10 (10 March 1924), 254-256.

At the armistice in November 1918, the Naval Aircraft Factory (authorized in July 1917) was producing one twin-engine flying boat per week. The factory, which extended over 40 acres along the Delaware, employed 3,600 people (including about 900 women, who received equal pay for equal work) and provided work to 6,000 others in industry. Of the 3,600 employees, only 25 had previous experience working on aircraft, requiring the factory to organize an apprentice school. Prior to 1921, the factory's work was almost entirely quantity production. Beginning in 1921, however, the factory focused on experimental and developmental projects, projects that required "as much engineering or design work as if the order were for a hundred." By 1925, the Navy's two huge plants at the factory, under the same industry attacks of "unfair competition" as McCook Field, tried to justify its activities. In 1925, the Navy's factory and flying field occupied 202 acres. "Navy's Part in Aviation Summarized," <u>Air Service Journal</u>, III:24 (14 Dec. 1918), 9, 13-14. "The Naval Aircraft Factory," <u>Aviation</u>, VI:1 (1 Feb. 1919), 28-30. Lt. Cmdr. S.J. Ziegler, Jr, Prod. Superintendent, Naval Ac. Factory, USN, "The Naval Aircraft Factory," <u>Aero Digest</u>, VII:3 (Sep. 1925), 465-469.

[147]   Eng. Div. Rpt. Nr. 238.

[148]   Eng. Div. Rpt. Nr. 238.

[149]   Maj. John F. Curry, Cmdr, Eng. Div, AS, "What McCook Field Means to Aviation," <u>U.S. Air Services</u>, XI:8 (Aug. 1926), 43-44. Curry's article, prepared in response to a request by the Dayton newspapers, had been tentatively entitled "What McCook Field Means to Dayton," but his assistant cautioned him against that title because it could be interpreted as "urging the retention of McCook Field in Dayton as a means of exploiting Dayton." Actually, however, congress already appropriated funds by then for building Wright

Field. Memo, Capt. O.S. Ferson, Asst, to Maj. J.F. Curry, Cmdr, Eng. Div, AS, subj: [McCook Field Article], 11 Dec. 1925.

The best single source for understanding McCook Field's organization and functions after General Patrick's redirection of acquisition policies is the Eng. Div. Rpt. Nr. 238.

McCook Field published a popularized account of the Engineering Division's activities as they progressed by 1924 in a 32-page booklet under the title A Little Journey to the Home of the Engineering Division, Army Air Service, n.d. [c. 1924]. Numerous photographs illustrate the field's activities and technical accomplishments as government support for the division was declining. Maurer article, 21-33; and Worman article, 12-15, 34-36.

150   Eng. Div. Rpt. atch. to ltr, Maj. W.G. Kilner, Exec, AS, to Ch, Eng. Div, AS, n.s. [Report on Research and Development Progress], 18 Feb. 1925, in Purtee, Appendix G. Walker-Wickam, 196. Wilson, Autobiography, 34.

Although the Engineering Division had no formal reciprocity agreements covering patents with industry, the division did hold contractors harmless from liability occurring because of patent infringements on experimental work and on some production work. The division agreed to let contractors use certain patents which the firms did not have and agreed to assume the infringement risks in return for the contractors using their skills and efforts in the performance of contracts and any patented features the companies held or controlled. In this way, the division was able to obtain information from industry's aeronautical experts. Eng. Div. Rpt. atch. to ltr, Maj. W.G. Kilner, Exec, AS, to Ch, Eng. Div, AS, n.s. [Report on Research and Development Progress], 18 Feb. 1925, in Purtee, Appendix G.

151   "A Slam at Our Friend -- The Enemy," Slipstream, V:12 (Dec. 1924), 26. Eng. Div. Rpt. Nr. 238. T.C. McMahon, "Something About McCook Field," Aero Digest, VI:4 (April 1925), 193-196. A more balanced assessment of McCook Field's role in U.S. aviation came from a Frenchman, Louis Breguet, "My Impressions of Aviation in America," Slipstream, VII:2 (Feb. 1926), 7-9.

152   C.G. Grey, "Impression of Aviation in America," Aviation, XVII:14 (6 Oct. 1924), 1082-1083. This was Grey's first article in a series that continued through 1924 evaluating aviation in the U.S.

153   C.G. Grey, "Impression of Aviation in America," Aviation, XVII:14 (6 Oct. 1924), 1082-1083.

154   C.G. Grey, "Post Impressions of American Aviation," Aviation, XVII:24 (15 Dec. 1924), 1395-1397.

155   C.G. Grey, "Post Impressions of American Aviation," Aviation, XVII:24 (15 Dec. 1924), 1395-1397.

156   C.G. Grey, "Post Impressions of American Aviation," Aviation, XVII:24 (15 Dec. 1924), 1395-1397.

157   C.G. Grey, "Post Impressions of American Aviation," Aviation, XVII:24 (15 Dec. 1924), 1395-1397.

158   "Fostering Development," Aviation, XVIII:21 (25 May 1925), 569.

159   "Aeronautical Engineering Indigestion," Aviation, XIX:25 (21 Dec. 1925), 869.

160   Purtee, 149-152. "Army and Navy Procurement," Aircraft Year Book 1934, 122.

161   "Army and Navy Procurement," Aircraft Year Book 1934, 122-123, 125-126. Wilson, Air Power, 34. Wilson, Autobiography, 72. "The Army Air Bill," Aviation, XXI:2 (12 July 1926), 53-54. Mooney & Layman, 78-80. Goldberg, 36.

The situation in the U.S. prior to the legislative windfall of 1926 was best described by the editor of the English publication, The Aeroplane, and quoted in C.G. Grey, "Air Legislation in the United States," Aviation, XVII:21 (24 Nov. 1924), 1302. Wrote Grey:

> One of the most amazing and amusing things about American aviation to any visitor from Europe is the fact that anybody in America can put together a few bits of stick and fabric and an engine, persuade it into the air by hoping for the best, call it an airplane, and proceed forth to "ply for hire or reward" without fee, license, registration, airworthiness certificate or anything else to show that the owner and pilot are competent to fly or are worthy to be trusted with the lives of their fellow-citizens.

The legislation of 1926 stemmed from Mitchell's final storm, this one over the Shenandoah disaster. One reaction to the controversy and his subsequent court-martial was President Coolidge's appointment in mid-September 1925 of the 9-member Morrow board to investigate U.S. aviation. Prior to this board, more than 20 congressional and other government committees conducted inquiries into aviation -- usually without much result. The president's board, led by financier Dwight W. Morrow, finished its investigation in mid-October before the 9-member Lampert committee completed a similar investigation for the House of Representatives.

The Morrow board submitted its unanimous report on 2 December 1925. Most of its recommendations were enacted into legislation in 1926: five-year plans and appropriations for the military; an end to competition with industry by the government; recognition of proprietary rights in designs; limits on competitive bidding to stop ruinous price cutting; and an aviation bureau in the Department of Commerce to assist and regulate civil aviation. The legislation featured the Air Commerce Act of 1926, approved by President Coolidge on 20 May, which brought commercial aviation under federal regulation. The act provided an assistant secretary of commerce for aviation and more consistent national control and support of commercial aviation. Congress soon voted the first money for civil aviation. Legislation also amended the Kelly Act of 1925, allowing the Post Office Department to contract for private control of air mail; in turn, the Postmaster General turned over all government-operated airlines to private contractors.

The commerce act was followed by the Air Corps Act, approved on 2 July 1926, establishing the Army Air Corps in place of the Air Service, providing five-year programs for both the Air Corps and the Navy's Bureau of Aeronautics, establishing assistant secretaries for aviation in both the War and Navy Departments, and revising military procurement. The Air Corps Act did not give military aviation independent status, either as a separate department or an autonomous branch of the War Department. But retitling of the Air Service as the Air Corps (modeled after the Navy's Marine Corps), implied that it was capable of dual operation: first, as an auxiliary and supporting force; and second, as an independent force operating alone on a separate mission. The administrative staff of the new corps was to consist of a chief of the Air Corps with the rank of major general, three assistants with the rank of brigadier general (rather than only one), 1,514 officers in grades between colonel and second lieutenant, and 16,000 enlisted. A five-year program of expansion was to be undertaken. The Air Corps was, however, to be under the immediate supervision of a new Assistant Secretary of War for Air, and the budget was to be entirely managed from the office of the Secretary of War. General Patrick regarded the Air Corps Act as an important step in the right direction, but that was still short of an independent air force. Implementation generally went forward in 1927.

Earlier, in January 1926, Daniel Guggenheim established the Daniel Guggenheim Fund for the Promotion of Aeronautics with the provision of $2.5 million to support aviation research. The fund closed in 1930 after providing seed money for many areas of aviation progress, including research centers, educational institutions, air law, aviation education, and air transportation.

Aircraft Year Book 1927, xii, 1-19, 70, 72-74, 83-89; these legislative milestones are reproduced on pages 346-372. "President's Air Board Reports," Aviation, XIX:24 (14 Dec. 1925), 834-837. "The Lampert

Committee Report," Aviation, XIX:26 (28 Dec. 1925), 906-909. "The Army Air Bill," Aviation, XXI:2 (12 July 1926), 53-54. "The Navy Air Bill," Aviation, XXI:3 (19 July 1926), 86. Loening, Takeoff, 162, 172-175, 177, 179-180, 182. Hallion, Guggenheim Contribution, 13-14, 31, 84-85. Wilson, Air Power, 26-36, 168-169. Wilson, Autobiography, 60-63, 68-72, 95-96.

Manufacturers regarded the legislation of 1926 as a Magna Carta, making the airplane industry "a business instead of an adventure." The law provided assurance for aircraft designers of winning models that they would also be allowed to manufacture them and not have to compete with firms bidding only to produce the designs. The benefits to commercial aviation of dependable military purchases of aircraft also were recognized. C.M. Keys, President, "Summary of Business for 1926, Curtiss Aeroplane & Motor Company, Inc.," Slipstream, VIII:4 (April 1927), 28-29. Edward P. Warner, Asst. Secy/Navy/Aeronautics, "Maritime Aviation," Slipstream, VIII:5 (May 1927), 10-11. Capt. Burdette S. Wright, "Progress of the Air Corps Under Administration of Maj. Gen. Mason M. Patrick," Slipstream, VIII:12 (Dec. 1927), 12, 16, 18, 22, 23; in that same issue, see also, "Other Accomplishments of the Air Corps," 23, 26; and "Year's Review in Development of Commercial Aviation," 26-28. Wilson, Autobiography, 47-48.

The 1926 legislation was regarded as the means by which the U.S. would catch up to European commercial aviation. But at least one major U.S. periodical challenged the common idea that American aviation was lagging far behind Europe's. The reason, asserted the editor, was that "no ideal type of airliner has as yet been built." Moreover, he indicated, "We have the engines." He did concede that the U.S. trailed Europe in operating experience. "Air Transportation," Aviation, XIX:23 (7 Dec. 1925), 813-814.

Actually, however, the U.S. already gained considerable experience. The Air Service had begun flying air mail for the Post Office Department during WWI, starting operations from Washington to New York on 15 May 1918 using Curtiss JN-4H aircraft with Hispano engines. The point of the operation was to give military pilots training, but within three months the air mail was transferred to the Post Office Department. The last flight of the Post Office operation occurred nine years later, on September 9, 1927. Under the Air Mail Law of 1925, known as the Kelly Act, the government entrusted delivery of air mail to private contractors. Growth of commercial aviation was also stimulated by this indirect subsidy. Loening, Takeoff, 138-143.

Meanwhile, in February 1921, the Army inaugurated a model airway between Washington and Dayton. The airway operated on a regular schedule and was gradually extended. By 1923, Army and Navy aircraft were flying between 300,000 and 400,000 miles a year. The airway provided experience for making transportation a reality, trained pilots in cross-country flying under strict regulations and scheduling, and tested new navigation instruments. A 12-seat version of the two-engine Liberty Martin bomber was developed for the airway. Loening, Takeoff, 148.

162   Goldberg, 36-37. W. Frank James, "What the Sixty-Ninth Congress Has Done for the Air Corps," Aero Digest, X:4 (April 1927), 274, 347-348. Aircraft Year Book 1928, 121-123.

163   Purtee, 156, 159. Walker-Wickam, 104, 127. In addition, the chief of the Materiel Division also directed the Air Corps Engineering School (now the Air Force Institute of Technology) and the Army Aeronautical Museum at Wright Field, both of which had their origin at McCook Field.

164   Goldberg, 37. "McCook Field, Dayton, Ohio, October 8th," Air Corps News Letter, X:14 (16 Oct. 1926), 20. Brig. Gen. William E. Gillmore, Ch, Matl. Div, AAC, "Review of Year's Developments in the Army Air Corps," Slipstream, VIII:3 (March 1927), 7. Aircraft Year Book 1928, 130-132.

Apparently, General Gillmore was originally ordered to establish his temporary headquarters at Wilbur Wright Field, but his headquarters seem to have been at McCook Field. See "Dayton Gets Supply

Division -- Headquarters Now at Wilbur Wright Field," Slipstream, VII:10 (Oct. 1926), 14; and "America's Greatest Air Station," Slipstream, VIII:8 (Aug. 1927), 14, 26.

[165] Walker-Wickam, 130.

[166] Purtee, 115-116. Frey, 112-113, 189-190. Claussen, 16.

As late as March 1918, assuming the minutes of the Aircraft Board as presented by McFarland (page 1134) are accurate, Deeds still regarded McCook Field as temporary until Langley became operational:

At the meeting of March 21, 1918, it was recommended that the board pass a resolution with regard to the construction of a propeller testing laboratory at McCook Field; and at the same time it was urged by Colonel Deeds that consideration should be given to the fact that this field was for war purposes only and that such laboratory should be designed to fit into the permanent organization at Langley Field after the war. However, on March 26th such a resolution was passed recommending to the Chief Signal Officer that he cause such a laboratory to be constructed and equipped with necessary apparatus and housing facilities at a cost of not more than $80,000.

[167] Walker-Wickam, 104-105. "Air Service Fields Purchased," DMA Weekly News Letter, I:n.nr. (22 March 1919), 13-14. "Government May Buy Old Dayton Field," Air Service Journal, IV:13 (29 March 1919), 1. "Do You Remember," Slipstream, I:7 (15 Dec. 1919), 19, 29. R.I. Markey, "The Spotlight on McCook Field," The Ohio State Engineer (Nov. 1920), 11. "The Present and Future Situation of the Engineering Division, Air Service, Dayton, Ohio," Slipstream, IV:8 (Aug. 1923), 19-22.

[168] Walker-Wickam, 105-107. "General Menoher's Annual Report," Aviation, IX:14 (20 Dec. 1920), 448-452. Maj. T.H. Bane, Ch, Eng. Div, AS, "A Message from the Commanding Officer," Slipstream, II:12 (Christmas 1920), 1. Maj. T.H. Bane, Ch, Eng. Div, AS, "A Message from the Commanding Officer," Slipstream, III:21 (Dec. 1921), 2.

[169] Walker-Wickam, 107-108. H.W. Karr, "Dayton, Ohio: As the Center of All Future U.S. Air Service Work," Slipstream, IV:9 (Sep. 1923), 5.

[170] Walker-Wickam, 108-111. "Proposed New Location of the Air Service Engineering Division," Slipstream IV:2 (Feb. 1923), 12-13. H.W. Karr, "Dayton, Ohio: As the Center of All Future U.S. Air Service Work," Slipstream, IV:9 (Sep. 1923), 5-7. "Wilbur Wright Memorial," Aero Digest, V:4 (Oct. 1024), 216-218. "'Greater McCook Field' Remains Near Dayton," Slipstream, VI:1 (Jan. 1925), 21-22.

[171] Maj. Thurman H. Bane, Ch, Eng. Div, AS, "Notes from Speech by Major Bane," Slipstream, IV:1 (Jan. 1923), 44.

[172] Fred F. Marshall, "Who Will Get McCook Field?" Slipstream, V:6 (June 1924), 5-6. "Publisher's News Letter," and "Slipstream's Reply to Aviation Weekly," Slipstream, V:7 (July 1924), 12-14.

[173] Purtee, 155-156. Maj. Gen. Mason M. Patrick, Ch, AAC, "Wright Field -- Pride of the Air Corps," Slipstream, VIII:10 (October 1927), 18.

[174] Carr article. "Dayton Gives Army Air Service Large Flying Field," U.S. Air Services, IX:9 (Sep. 1924), 23. "Wilbur Wright Memorial," Aero Digest, V:4 (Oct. 1924), 216-218. "McCook on New Site Assured for Dayton," Slipstream, V:11 (Nov. 1924), 33. "Dayton Is Now Undisputed Air Center," Slipstream, VIII:2 (Feb. 1927), 13. "World's Most Widely Known Air Station Being Razed," Slipstream, VIII:5 (May 1927), insert. "Air Corps Laboratories Open Oct. 12.," Aviation, XXIII:14 (3 Oct. 1927), 839-840. "Wright Field Is Dedicated," Aviation, XXIII:17 (24 Oct. 1927), 995-997. "Dedication of Wright Field," U.S. Air Services, XII:11 (Nov. 1927), 32. Walker-Wickam, 112-115, 118-122.

The last day of government flight operations at McCook Field was 30 June 1927. By June, aircraft practically stopped going to McCook Field, opting for Wright Field instead. "No Landings at McCook Field After June 30th," Air Corps News Letter, XI:8 (27 June 1927), 184.

Three activities of the Materiel Division continued to operate at McCook Field until 1929: the dynamometer laboratory, the propeller test building and test stands, and the 5-foot wind tunnel.

Wilbur Wright Field was an important training center in WWI and was used for testing, especially for those operations that McCook Field was too small to handle. On 21 August 1925, the War Department changed the name of the site to Wright Field. Six years later, on 1 July 1931, the part of Wright Field east of Huffman Dam was renamed Patterson Field. The two fields were combined and became Wright-Patterson Air Force Base on 13 January 1948. "Name of New Field for Engineering Division," Air Service News Letter, IX:14 (6 Aug. 1925), 3. Dean, 150. Termena, Peiffer, and Carlin, 6.

After additional purchases, Wright Field comprised two unequal parts: 750 acres beyond Huffman Dam (now Area B of Wright-Patterson AFB), the site of the Materiel Division's experimental facilities and flying field; and 3,800 acres in the flood-control basin of the Mad River (now Areas A and C of WPAFB.) The latter site was earlier used by the Wright brothers for their experimental flying and their flying school.

Others present at the Wright Field dedication ceremonies included Secretary of War Dwight F. Davis, Assistant Secretary of War F. Trubee Davison, Air Corps Chief Major General Mason M. Patrick, and Dayton industrialist Edward A. Deeds, former chief of the Equipment Division.

Other sources include "Government Accepts New McCook Field Site at Dayton, Ohio," Slipstream V:9 (Sep. 1924), 7-9; A.M. Jacobs, "The New Wright Field," Air Corps News Letter, XI:11 (30 Aug. 1927), 251-252; Maj. Gen. Mason M. Patrick, Ch, AAC, "Wright Field -- Pride of the Air Corps," Slipstream, VIII:10 (October 1927), 18-20, 48; "Wright Field Dedicated," Slipstream, VIII:11 (Nov. 1927), 17, 32.

Interestingly, the move to Wright Field coincided with the completion of a three-month tour of the 48 states by 25-year old Charles Augustus Lindbergh. Starting on 20 July 1927, Lindbergh covered 22,350 miles in 260 hours flying time without an accident. He made 82 stops and was late only once during the tour that ended on 23 October. An estimated 50 million people saw the aircraft and attended the receptions at the various stops. The tour, conducted to promote commercial aviation under the auspices of the Daniel Guggenheim Fund for the Promotion of Aeronautics and the Department of Commerce, was considered a greater practical feat than Lindbergh's triumphal 33-hour, 3,610-mile flight across the Atlantic from New York to Paris on 20 and 21 May. Lindbergh's aircraft, a Ryan monoplane, was powered by a single Wright air-cooled J-5C Whirlwind engine designed by Charles L. Lawrence. The engine was also the product of work by Jones, Heron, and Morehouse -- all who came from McCook Field in 1926. The aircraft also used a duralumin propeller made by Standard Steel Propeller. Lindbergh's Atlantic flight, coupled with others in 1927, did for military aviation in the U.S. what Bleriot's flight across the English channel had accomplished for Europe: reduced national security by opening the possibility of eventual bombing of U.S. industrial centers.

Capt. Charles A. Lindbergh, "My Flight to Paris," Aero Digest, X:6 (June 1927), 514, 516, 518, 520, 522, 524. "Lindbergh Conquers the World by Airplane," U.S. Air Services, XII:6 (June 1927), 17-23. "Technical Notes on Trans-Atlantic Plane," Slipstream, VIII:6 (June 1927), 31. An entire issue of Slipstream, VII:7 (July 1927), was dedicated to Lindbergh's famous flight. Maj. Gen. Mason M. Patrick, "Aviation's Future in United States," Aero Digest, XI:1 (July 1927), 41, 110. "Lindbergh on Tour," U.S. Air Services, XII:8 (Aug. 1927), 56, 58. "Lindbergh Completes Tour," Slipstream, VIII:11 (Nov. 1927), 32. "The Lindbergh Tour," U.S. Air Services, XII:12 (Dec. 1927), 39-40. Aircraft Year Book 1928, 3-19,

37-38, 47, 49. Loening, Takeoff, 195-196. Wilson, Autobiography, 102-108. Heron, Autobiography, 142-147.

The numerous record flights of 1927 were symbolic of the year's pivotal position. The cumulative impact of these flights on public awareness of aviation progress and safety gave a great boost to commercial aviation, as did air mail. Students flocked to flying schools. The demand for commercial services led to increased aircraft manufacturing. Lindbergh embarked on another tour on 13 December 1927, flying 2,100 miles non-stop from Washington, D.C., to Mexico City in 27 hours. From there, he traveled to Central and South America. Aircraft Year Book 1928, 38-41, 49, 51, 117-120.

Another transoceanic flight, this one the first across the Pacific, was made by Lieutenants Lester J. Maitland and Albert F. Hegenberger. They flew from Oakland, California, to Hawaii starting on 28 June. Their 26-hour, 2,400-mile flight was made in a C-2 Fokker equipped with three air-cooled Wright Whirlwind engines of 200-horsepower. The aircraft and equipment were tested at Wright Field, and on the flight to California. "Hawaiian Flight a Remarkable Achievement," Air Corps News Letter, XI:9 (19 July 1927), 205-207. "The Army Flight to Hawaii," Aero Digest, XI:1 (July 1927), 16-18. Frederick R. Neely, "Army Makes Longest Over-Water Flight," U.S. Air Services, XII:8 (Aug. 1927), 18-22. Aircraft Year Book 1928, 22-24.

[175] Maj. Gen. Mason M. Patrick, Ch, AAC, "Wright Field -- Pride of the Air Corps," Slipstream, VIII:10 (October 1927), 18.

The chief of the Engineering Division characterized the 10 years prior to 1926 as the decade of the military aircraft. He expected commercial aviation to dominate the next decade. See Maj. John F. Curry, Ch, Eng. Div, AAC, "An Outline of Aeronautical Development," Aviation, XXI:5 (2 Aug. 1926), 175.

www.ingramcontent.com/pod-product-compliance
Lightning Source LLC
Chambersburg PA
CBHW080809180526
45168CB00006B/2384